Jutta Kühn

# AlGaN/GaN-HEMT Power Amplifiers with Optimized Power-Added Efficiency for X-Band Applications

Karlsruher Forschungsberichte
aus dem Institut für Hochfrequenztechnik und Elektronik

*Herausgeber: Prof. Dr.-Ing. Thomas Zwick*

**Band 62**

# AlGaN/GaN-HEMT Power Amplifiers with Optimized Power-Added Efficiency for X-Band Applications

von
Jutta Kühn

KIT Scientific Publishing

Dissertation, Karlsruher Institut für Technologie
Fakultät für Elektrotechnik und Informationstechnik, 2010

**Impressum**

Karlsruher Institut für Technologie (KIT)
KIT Scientific Publishing
Straße am Forum 2
D-76131 Karlsruhe
www.ksp.kit.edu

KIT – Universität des Landes Baden-Württemberg und nationales
Forschungszentrum in der Helmholtz-Gemeinschaft



KIT Scientific Publishing 2011
Print on Demand

ISSN 1868-4696
ISBN 978-3-86644-615-1

**Vorwort des Herausgebers**

Solid state amplifiers are the key components of modern radar and wireless communication systems. The performance of the overall systems, especially at the transmitting side, is limited by the available output power of the last amplifier in the chain/line. This becomes more evident when the operation moves towards millimeter wave frequencies. The reason for this limitation is due to the scaling down of semiconductor devices in order to achieve higher operating frequencies but at the same time compromises the power handling capability. As such, the use of new semiconductor materials becomes the focus of recent worldwide research. The current most promising technology is the gallium nitride (GaN), which offers high power density, high efficiency and exceptional wideband performance as compared to existing technologies such as silicon (SI) or gallium arsenide (GaAs). A key advantage to gallium nitride is its ability to maintain high operating voltages at higher frequency ranges. This is due to the unique combination of high bandgap energy resulting in higher device tolerance for high maximum field strengths, as well as the high speed properties of electron mobility and saturation velocity. These advantages thus distinguish the GaN from other competing technologies in terms of power generation in operating frequencies of 10 GHz and above. Besides the available output power the efficiency plays an important role in any device with limited energy and/or heat dissipation, which is the case for nearly all systems. Since up to now the power added efficiencies of GaN devices at GHz frequencies are still below what GaAs can offer, this efficiency factor becomes an important research task for the scientific community.

The dissertation of Dr. Jutta Kühn focuses on the design of AlGaN/GaN HEMT high-power amplifiers in Class-AB operation for X-band (8-12GHz) applications with particular emphasis on an optimized power added efficiency (PAE). In this work a PAE enhancement at the device level with improved design techniques achieved output power levels of several watts and PAEs above 50%. The presented work of Ms. Kühn is a fundamental basis for future research in the field of AlGaN/GaN-HEMT Power Amplifiers. The results will draw attention worldwide and entail more relevant research. I wish her all the best for her future.

Prof. Dr.-Ing. Thomas Zwick
- director of the IHE -

## Forschungsberichte aus dem Institut für Höchstfrequenztechnik und Elektronik (IHE) der Universität Karlsruhe (TH) (ISSN 0942-2935)

---

Herausgeber: Prof. Dr.-Ing. Dr. h.c. Dr.-Ing. E.h. Werner Wiesbeck

Band 1 Daniel Kähny
**Modellierung und meßtechnische Verifikation polarimetrischer, mono- und bistatischer Radarsignaturen und deren Klassifizierung** (1992)

Band 2 Eberhardt Heidrich
**Theoretische und experimentelle Charakterisierung der polarimetrischen Strahlungs- und Streueigenschaften von Antennen** (1992)

Band 3 Thomas Kürner
**Charakterisierung digitaler Funksysteme mit einem breitbandigen Wellenausbreitungsmodell** (1993)

Band 4 Jürgen Kehrbeck
**Mikrowellen-Doppler-Sensor zur Geschwindigkeits- und Wegmessung - System-Modellierung und Verifikation** (1993)

Band 5 Christian Bornkessel
**Analyse und Optimierung der elektrodynamischen Eigenschaften von EMV-Absorberkammern durch numerische Feldberechnung** (1994)

Band 6 Rainer Speck
**Hochempfindliche Impedanzmessungen an Supraleiter / Festelektrolyt-Kontakten** (1994)

Band 7 Edward Pillai
**Derivation of Equivalent Circuits for Multilayer PCB and Chip Package Discontinuities Using Full Wave Models** (1995)

Band 8 Dieter J. Cichon
**Strahlenoptische Modellierung der Wellenausbreitung in urbanen Mikro- und Pikofunkzellen** (1994)

Band 9 Gerd Gottwald
**Numerische Analyse konformer Streifenleitungsantennen in mehrlagigen Zylindern mittels der Spektralbereichsmethode** (1995)

## Forschungsberichte aus dem Institut für Höchstfrequenztechnik und Elektronik (IHE) der Universität Karlsruhe (TH) (ISSN 0942-2935)

Band 10 Norbert Geng
**Modellierung der Ausbreitung elektromagnetischer Wellen in Funksystemen durch Lösung der parabolischen Approximation der Helmholtz-Gleichung** (1996)

Band 11 Torsten C. Becker
**Verfahren und Kriterien zur Planung von Gleichwellennetzen für den Digitalen Hörrundfunk DAB (Digital Audio Broadcasting)** (1996)

Band 12 Friedhelm Rostan
**Dual polarisierte Microstrip-Patch-Arrays für zukünftige satellitengestützte SAR-Systeme** (1996)

Band 13 Markus Demmler
**Vektorkorrigiertes Großsignal-Meßsystem zur nichtlinearen Charakterisierung von Mikrowellentransistoren** (1996)

Band 14 Andreas Froese
**Elektrochemisches Phasengrenzverhalten von Supraleitern** (1996)

Band 15 Jürgen v. Hagen
**Wide Band Electromagnetic Aperture Coupling to a Cavity: An Integral Representation Based Model** (1997)

Band 16 Ralf Pötzschke
**Nanostrukturierung von Festkörperflächen durch elektrochemische Metallphasenbildung** (1998)

Band 17 Jean Parlebas
**Numerische Berechnung mehrlagiger dualer planarer Antennen mit koplanarer Speisung** (1998)

Band 18 Frank Demmerle
**Bikonische Antenne mit mehrmodiger Anregung für den räumlichen Mehrfachzugriff (SDMA)** (1998)

Band 19 Eckard Steiger
**Modellierung der Ausbreitung in extrakorporalen Therapien eingesetzter Ultraschallimpulse hoher Intensität** (1998)

## Forschungsberichte aus dem Institut für Höchstfrequenztechnik und Elektronik (IHE) der Universität Karlsruhe (TH) (ISSN 0942-2935)

Band 20 Frederik Küchen
**Auf Wellenausbreitungsmodellen basierende Planung terrestrischer COFDM-Gleichwellennetze für den mobilen Empfang** (1998)

Band 21 Klaus Schmitt
**Dreidimensionale, interferometrische Radarverfahren im Nahbereich und ihre meßtechnische Verifikation** (1998)

Band 22 Frederik Küchen, Torsten C. Becker, Werner Wiesbeck
**Grundlagen und Anwendungen von Planungswerkzeugen für den digitalen terrestrischen Rundfunk** (1999)

Band 23 Thomas Zwick
**Die Modellierung von richtungsaufgelösten Mehrwegegebäudefunkkanälen durch markierte Poisson-Prozesse** (2000)

Band 24 Dirk Didascalou
**Ray-Optical Wave Propagation Modelling in Arbitrarily Shaped Tunnels** (2000)

Band 25 Hans Rudolf
**Increase of Information by Polarimetric Radar Systems** (2000)

Band 26 Martin Döttling
**Strahlenoptisches Wellenausbreitungsmodell und Systemstudien für den Satellitenmobilfunk** (2000)

Band 27 Jens Haala
**Analyse von Mikrowellenheizprozessen mittels selbstkonsistenter finiter Integrationsverfahren** (2000)

Band 28 Eberhard Gschwendtner
**Breitbandige Multifunktionsantennen für den konformen Einbau in Kraftfahrzeuge** (2001)

Band 29 Dietmar Löffler
**Breitbandige, zylinderkonforme Streifenleitungsantennen für den Einsatz in Kommunikation und Sensorik** (2001)

Band 30 Xuemin Huang
**Automatic Cell Planning for Mobile Network Design: Optimization Models and Algorithms** (2001)

## Forschungsberichte aus dem Institut für Höchstfrequenztechnik und Elektronik (IHE) der Universität Karlsruhe (TH) (ISSN 0942-2935)

---

Band 31 Martin Fritzsche
**Anwendung von Verfahren der Mustererkennung zur Detektion von Landminen mit Georadaren** (2001)

Band 32 Siegfried Ginter
**Selbstkonsistente Modellierung der Erhitzung von biologischem Gewebe durch hochintensiven Ultraschall** (2002)

Band 33 Young Jin Park
**Applications of Photonic Bandgap Structures with Arbitrary Surface Impedance to Luneburg Lenses for Automotive Radar** (2002)

Band 34 Alexander Herschlein
**Entwicklung numerischer Verfahren zur Feldberechnung konformer Antennen auf Oberflächen höherer Ordnung** (2002)

Band 35 Ralph Schertlen
**Mikrowellenprozessierung nanotechnologischer Strukturen am Beispiel von Zeolithen** (2002)

Band 36 Jürgen von Hagen
**Numerical Algorithms for the Solution of Linear Systems of Equations Arising in Computational Electromagnetics** (2002)

Band 37 Ying Zhang
**Artificial Perfect Magnetic Conductor and its Application to Antennas** (2003)

Band 38 Thomas M. Schäfer
**Experimentelle und simulative Analyse der Funkwellenausbreitung in Kliniken** (2003)

Band 39 Christian Fischer
**Multistatisches Radar zur Lokalisierung von Objekten im Boden** (2003)

Band 40 Yan C. Venot
**Entwicklung und Integration eines Nahbereichsradarsensorsystems bei 76,5 GHz** (2004)

Band 41 Christian Waldschmidt
**Systemtheoretische und experimentelle Charakterisierung integ-**

## Forschungsberichte aus dem Institut für Höchstfrequenztechnik und Elektronik (IHE) der Universität Karlsruhe (TH) (ISSN 0942-2935)

---

Band 42 Marwan Younis
**Digital Beam-Forming for high Resolution Wide Swath Real and Synthetic Aperture Radar** (2004)

Band 43 Jürgen Maurer
**Strahlenoptisches Kanalmodell für die Fahrzeug-Fahrzeug-Funkkommunikation** (2005)

Band 44 Florian Pivit
**Multiband-Aperturantennen für Basisstationsanwendungen in rekonfigurierbaren Mobilfunksystemen** (2005)

Band 45 Sergey Sevskiy
**Multidirektionale logarithmisch-periodische Indoor-Basisstationsantennen** (2006)

Band 46 Martin Fritz
**Entwurf einer breitbandigen Leistungsendstufe für den Mobilfunk in Low Temperature Cofired Ceramic** (2006)

Band 47 Christiane Kuhnert
**Systemanalyse von Mehrantennen-Frontends (MIMO)** (2006)

Band 48 Marco Liebler
**Modellierung der dynamischen Wechselwirkungen von hochintensiven Ultraschallfeldern mit Kavitationsblasen** (2006)

Band 49 Thomas Dreyer
**Systemmodellierung piezoelektrischer Sender zur Erzeugung hochintensiver Ultraschallimpulse für die medizinische Therapie** (2006)

Band 50 Stephan Schulteis
**Integration von Mehrantennensystemen in kleine mobile Geräte für multimediale Anwendungen** (2007)

Band 51 Werner Sörgel
**Charakterisierung von Antennen für die Ultra-Wideband-Technik** (2007)

Band 52 Reiner Lenz
**Hochpräzise, kalibrierte Transponder und Bodenempfänger für**

## Forschungsberichte aus dem Institut für Höchstfrequenztechnik und Elektronik (IHE) der Universität Karlsruhe (TH) (ISSN 0942-2935)

---

Band 53 Christoph Schwörer
**Monolithisch integrierte HEMT-basierende Frequenzvervielfacher und Mischer oberhalb 100 GHz** (2008)

Band 54 Karin Schuler
**Intelligente Antennensysteme für Kraftfahrzeug-Nahbereichs-Radar-Sensorik** (2007)

Band 55 Christian Römer
**Slotted waveguide structures in phased array antennas** (2008)

## Fortführung als "Karlsruher Forschungsberichte aus dem Institut für Hochfrequenztechnik und Elektronik" bei KIT Scientific Publishing (ISSN 1868-4696)

## Karlsruher Forschungsberichte aus dem Institut für Hochfrequenztechnik und Elektronik (ISSN 1868-4696)

Herausgeber: Prof. Dr.-Ing. Thomas Zwick

**Die Bände sind unter www.ksp.kit.edu als PDF frei verfügbar oder als Druckausgabe bestellbar.**

Band 55 Sandra Knörzer
**Funkkanalmodellierung für OFDM-Kommunikationssysteme bei Hochgeschwindigkeitszügen** (2009)
ISBN 978-3-86644-361-7

Band 56 Fügen, Thomas
**Richtungsaufgelöste Kanalmodellierung und Systemstudien für Mehrantennensysteme in urbanen Gebieten** (2009)
ISBN 978-3-86644-420-1

Band 57 Pancera, Elena
**Strategies for Time Domain Characterization of UWB Components and Systems** (2009)
ISBN 978-3-86644-417-1

Band 58 Timmermann, Jens
**Systemanalyse und Optimierung der Ultrabreitband-Übertragung** (2010)
ISBN 978-3-86644-460-7

Band 59 Pontes, Juan
**Analysis and Design of Multiple Element Antennas for Urban Communication** (2010)
ISBN 978-3-86644-513-0

Band 60 Lambrecht, Andreas
**True-Time-Delay Beamforming für ultrabreitbandige Systeme hoher Leistung** (2010)
ISBN 978-3-86644-522-2

## Karlsruher Forschungsberichte aus dem Institut für Hochfrequenztechnik und Elektronik (ISSN 1868-4696)

---

Band 61 Adamiuk, Grzegorz
**Methoden zur Realisierung von dual-orthogonal, linear polarisierten Antennen für die UWB-Technik** (2010)
ISBN 978-3-86644-573-4

Band 62 Kühn, Jutta
**AlGaN/GaN-HEMT Power Amplifiers with Optimized Power-Added Efficiency for X-Band Applications** (2011)
ISBN 978-3-86644-615-1

# AlGaN / GaN-HEMT Power Amplifiers with Optimized Power-Added Efficiency for X-Band Applications

Zur Erlangung des akademischen Grades eines

**DOKTOR-INGENIEURS**

von der Fakultät für
Elektrotechnik und Informationstechnik,
Karlsruher Institut für Technologie (KIT)

genehmigte

**DISSERTATION**

von

**Dipl.-Ing. Jutta Kühn**

geboren in Freiburg, Deutschland

Tag der mündlichen Prüfung: 06.12.2010
Hauptreferent: Prof. Dr. rer. nat. Dr. h.c. Manfred Thumm
Korreferent: Prof. Dr. rer. nat. Oliver Ambacher

*To Jochen, Eva*

*&*

*my parents*

# Contents

# Kurzfassung

Leistungsverstärker sind Bestandteil eines jeden drahtlosen mehrfunktionalen Hochfrequenzsystems. Da die Verstärker einen sehr hohen Energieverbrauch haben und gleichzeitig die Primärenergie autarker Systeme limitiert ist, ist die Entwicklung hocheffizienter Leistungsverstärker für autarke Kommunikations- und Radarsysteme wie z.B. für phasengesteuerte Array-Antennen unabdingbar, um zusätzliche Kühlungsenergie einzusparen. Unter den vorhandenen Halbleiterbauelementtechnologien ist Galliumnitrid (GaN) aufgrund seiner herausragenden Materialeigenschaften wie der mehr als 10-fachen Leistungsdichte verglichen mit Galliumarsenid (GaAs) sehr interessant für die genannten Anwendungsgebiete. Die hohe Leistungsdichte ermöglicht die Erzeugung hoher Leistungen von über 20 W im X-Frequenzband (8-12 GHz) auf Schaltungsebene, wodurch zukünftige Anforderungen nach hohen Leistungen über große Bandbreiten bei moderater Kompression erfüllt werden können.

Zu Beginn dieser Arbeit im Jahr 2006 lag der erzielbare Wirkungsgrad mit der relativ jungen noch nicht ausgereiften Technologie der GaN-Leistungstransistoren noch deutlich unter dem von Transistoren auf Basis anderer III-V Halbleiter wie z.B. GaAs. Auch die Integration des neuen GaN-basierten Hochfrequenz-Leistungselements und die effiziente Ausnutzung des hohen Impedanzniveaus von GaN-Transistoren innerhalb eines Schaltungsentwurfs wurde bis dahin nicht hinreichend untersucht. Während die ersten GaN-basierten Leistungsverstärker dieser Arbeit Wirkungsgrade im Bereich von 20-25 % bei Ausgangsleistungen von 15 W über eine Frequenzbandbreite von 1-2 GHz im X-Band aufwiesen, waren zeitgleich mit GaAs-basierten Verstärkern Wirkungsgrade von über 40 % bei jedoch kleinen Ausgangsleistungen um die 5 W im X-Band auf Schaltungsebene möglich.

Diese Arbeit ist aus der starken Nachfrage nach höheren Wirkungsgraden bei Leistungsverstärkern mit AlGaN/GaN High Electron Mobility Transistoren (HEMTs) entstanden. Zur Optimierung des Wirkungsgrades wurden verschiedene Ansätze und Konzepte auf Bauelement- und Schaltungsebene untersucht. Struktur- und Layoutveränderungen des Transistors und die Auswirkungen auf den erzielbaren Wirkungsgrad wurden analysiert, optimiert und in Leistungsverstärkerschaltungen integriert.

Neben der notwendigen Verbesserung des Wirkungsgrades befasst sich ein weiterer Teilaspekt dieser Arbeit mit der schlechten Ausgangsanpassung einstufiger AlGaN/GaN HEMT Hochleistungsverstärker, die für maximale Ausgangsleistung/ Effizienz ausgelegt sind. Vor allem für Anwendungen, bei denen starke Eingangs- oder Ausgangsfehlanpassungen auftreten können, ist eine optimierte Kleinsignalanpassung unabdingbar. Diese Arbeit liefert eine Lösung zur Verbesserung der Kleinsignalanpassung mit Hilfe eines balancierten Verstärkeraufbaus bei nur geringer Wirkungsgradeinbuße. Auf diese Wei-se wurde der erste Demonstrator für einen GaN-basierten balancierten Hochleistungsverstärker in Mikrostreifentechnologie für Anwendungen im X-Band entworfen und entwickelt.

Die Hauptaufgabe dieser Arbeit besteht jedoch im Entwurf innovativer Schaltungstechniken zur Optimierung des Wirkungsgrades GaN-basierter Hochleistungsverstärker. Die Anwendung bekannter Entwurfmaßnahmen GaAs-basierter Verstärker auf GaN-basierte Leistungsverstärker wurde in dieser Arbeit unter Beachtung der elektrischen Eigenschaften der GaN-Leistungstransistoren wie z.B. dem hohen Impedanzniveau, der Hochfrequenzdispersion, dem Gewinn und der Leistungskompression untersucht.

Die Arbeit konzentriert sich dabei auf zwei effektive Maßnahmen zur Optimierung des Wirkungsgrades. Auf der einen Seite wird der Einfluss des Leistungs- zu Treiberstufen Gateweitenverhältnisses auf das Wirkungsgradverhalten zweistufiger Leistungsverstärker mittels vereinfachender Annahmen untersucht. Anschließend wird die Theorie durch reale monolithisch integrierte AlGaN/GaN HEMT Mikrowellenverstärkerschaltungen mit unterschiedlichen Leistungs- zu Treiberstufen Gateweitenverhältnissen verifiziert. Veröffentlichungen in diesem Bereich weisen üblicherweise Leistungs- zu Treiberstufen Gateweitenverhältnisse von 2:1 - 3:1 auf. Die Studie hier zeigt jedoch, dass eine Erhöhung dieses Verhältnisses auf bis zu 4:1 eine zusätzliche Verbesserung um einige Wirkungsgradprozente zu Lasten der maximal möglichen Frequenzbandbreite bewirken kann, so dass anhand der Verstärkerspezifikationen ein Kompromiss zwischen erreichbarem Wirkungsgrad und möglicher Bandbreite geschlossen werden muss. Auf der anderen Seite behandelt diese Arbeit den Einfluss der Harmonischen Abschlüsse am Eingang und Ausgang eines AlGaN/GaN-Transistors anhand von Load-Pull Simulationen und der Realisierung geeigneter Testschaltungen. Da die zweite Harmonische das Wirkungsgradverhalten eines in Klasse-AB bis Klasse-B betriebenen Transistors sehr stark beeinflusst, konzentriert sich diese Untersuchung ausschließlich auf die Terminierung der zweiten Harmonischen. Der prinzipielle Wirkungsmechanismus der Terminierung der zweiten Harmonischen am Eingang sowie am Ausgang eines AlGaN/GaN-Transistors wurde nur unzureichend in der Literatur untersucht und wurde deshalb grundlegend in dieser Arbeit anhand von Load-Pull Simulationen eines AlGaN/GaN HEMT Modells theoretisch analysiert. Anschließend wurde die in der Theorie erwartete Wirkungsgradverbesserung durch die Terminierung der zweiten Harmonischen durch reale einstufige monolithisch integrierte Mikrowellenleistungsverstärkerschaltungen verifiziert. Es wird gezeigt, dass auch die Terminierung der zweiten Harmonischen eine vielversprechende Schaltungstechnik zur Verbesserung des Wirkungsgrades von GaN-Leistungsverstärkern unter Berücksichtigung einer reduzierten Bandbreite liefert.

Diese Arbeit bildet die Grundlage für den Entwurf GaN-basierter Hochleistungs-verstärker mit Wirkungsgraden von über 40 % bei gleichzeitigen Ausgangsleistungen von mehr als 20 W über bis zu 2 GHz Frequenzbandbreite. Damit ist der Wirkungsgrad GaN-basierter Leistungsverstärker heutzutage vergleichbar mit dem von z.B. Verstärkern auf GaAs-Basis bei gleichzeitig bis zu 5-facher Ausgangsleistung bzw. 3-facher Leistung pro Chip-fläche. Folglich bieten GaN-basierte Leistungsverstärker ein hohes Potenzial für Anwendungen zukünftiger monolithisch phasengesteuerter Array-Systeme.

# Abstract

High-power amplifiers (HPAs) are part of any advanced wireless multifunctional RF-system. Due to the high power consumption of the HPAs and due to the limited prime energy in autonomous systems, the development of highly-efficient power amplifiers is of great interest for modern communication and autonomous radar systems such as solid-state phased arrays as well as for airborne and space applications to save up additional cooling energy. Of the various semiconductors and device technologies currently available, the most promising material candidate for these applications is the wide bandgap gallium nitride (GaN). Due to its inherent and outstanding material properties such as a power density of more than a factor of ten compared to GaAs devices, GaN enables an impressive capability for high-power amplification resulting in a high output-power above 20 W on circuit level at X-band frequencies (8-12 GHz). Therefore, GaN-based HPAs have the potential to fulfill future demands on high output power levels over wide bandwidths at a moderate gain compression.

At the beginning of this work in 2006, the obtainable power-added efficiency (PAE) of the relatively young GaN technology was inferior compared to other existing III-V semiconductors such as GaAs due to the fact that the GaN-based transistor technology development was still in progress and had not reached maturity. In addition, the integration of the relatively new GaN RF power devices and the efficient use of the high-impedance level of the GaN devices within circuit designs is not thoroughly exploited and therefore plays a key role for the design of innovative high-power amplifiers targeting high efficiencies. While the first GaN-based HPAs within this work offered PAE values between 20-25 % with output power levels of about 15 W over frequency bandwidth of 1-2 GHz at X-band, GaAs-based HPAs showed PAE values above 40 % at low output power levels of about 5 W and comparable frequencies.

This work has arisen out of the strong demand for a superior power-added efficiency of AlGaN/GaN high electron mobility transistor (HEMT) HPAs on circuit level. Different concepts and approaches on device and design level for PAE improvements were analyzed. It is obvious that every single percent in PAE enhancement contributes a strong positive impact on systems that directly operate in the space or aircraft environment with limited primary energy for cooling. Structural and layout changes of the GaN transistor and the impact on the PAE behavior were analyzed, optimized, and finally integrated in advanced HPA monolithic microwave integrated circuits (MMICs).

Apart from the required PAE improvement, one important topic of this work is based on the poor output matching capability of single-ended AlGaN/GaN HEMT HPAs designed for high output power and/or efficiency. An improved small-signal matching capability is required for application fields where strong input and/or output mismatches occur, e.g. high power measurement equipment or X-frequen-cy-band near-field radar. This work presents an effective way to realize a highly improved small-signal matching capability at the expense of only a few percent in the PAE performance on the basis of a balanced amplifier. Thus, a first balanced microstrip HPA demonstrator in GaN technology for X-band frequencies was designed and developed.

However, the main aim of this work consists in advanced circuit design techniques for PAE improvements of GaN HEMT HPAs. The implementation of well-known design methods from GaAs-based PAs into GaN HEMT power amplifiers is analyzed in consideration of the electrical properties of GaN HEMTs such as the high impedance level, the RF dispersion, gain, and power compression of GaN devices.

Two design techniques are discussed in greater detail that both deliver an essential contribution to the required PAE enhancement. On the one hand, the impact of the power (PA)- to driver (DRV)-stage total gate-width (TGW) ratio on the PAE performance of dual-stage HPAs is fundamentally analyzed in theory and supported by real GaN-HEMT HPA MMICs offering different PA/DRV TGW ratios. According to the literature, commonly PA/DRV TGW ratios of about 2:1 - 3:1 are chosen for dual-stage HPA designs. Nevertheless, this study shows that higher PA/DRV TGW ratios of up to 4:1 may deliver an additional improvement of a few percent in PAE when skillfully used. However, the PA/DRV TGW ratio strongly impacts the trade-off between the obtainable PAE and possible frequency bandwidth. On the other hand, the influence of the harmonics at the input and output of a GaN HEMT device is determined by means of load-pull simulations and the realization of test HPA MMICs. Due to the strong impact of the second harmonic on the PAE behavior of a transistor in Class-AB to Class-B operation, this work focuses only on the termination of the second harmonic. The principal mechanism of second harmonic termination at both the input and output of a GaN-based transistor could not be satisfactorily found in literature and therefore is theoretically analyzed in detail within this work with help of load-pull simulations of an AlGaN/GAN HEMT model. Subsequently, the expected PAE improvement due to second harmonic termination from the theoretical assumptions is verified by real single-stage HPA MMIC designs. The study shows that the termination of the second harmonic also offers a promising design technique to improve the PAE of high-power amplifiers at the expense of a smaller frequency bandwidth.

Due to the above mentioned concepts and approaches on device and design level, this work provides the basis for the design of GaN-based high-power amplifiers with PAEs above 40 % and output power levels beyond 20 W over frequency bandwidths of up to 2 GHz. Thus, the PAE of present GaN-based HPA MMICs is comparable to that of e.g. GaAs-based HPA MMICs, but with an up to 5-times higher output power level or rather a 3-times higher power density per chip area at the same time. Therefore, GaN-based HPA MMICs are of great avail for future monolithic solid-state phased-array systems.

# Chapter 1

# Introduction

## 1.1 Motivation

Energy efficiency is a key requirement especially for airborne and space applications. At system levels, cooling, DC-DC conversion, and protective measures dominate to overall efficiency of RF systems. Monolithic Microwave Integrated Circuits (MMICs) based on the evolving gallium nitride (GaN) technology provide several advantages. For example, high operation bias close to on-board power supply of typical airborne and space systems saves additional energy for DC-DC conversion. In addition, due to its high robustness, GaN devices withstand changes in the operation bias, antenna mismatch, and are also less susceptible to radiation.

The development of high-power amplifiers (HPAs), being the main power-consumption part in any advanced wireless communication or airborne related system, is driven by the growing demand on marine, airborne and space-based radar systems for scientific research, commercial communication systems, and military applications [94], [95], [46], [119]. All these applications comprise a limited DC power supply that demands the availability of highly efficient HPAs.

In the near future, GaN technology development will cover all topics from the transistor device level to the complete modules for RF-applications. A first European transmit and receive (T/R)-module based on a GaN MMIC driver amplifier (DA), HPA, and low noise amplifier (LNA) is presented in [101]. An overview of the development activities with GaN RF-devices in Europe is given in [85].

Over the past few years, intensive research has been dedicated to the new GaN-based transistor technology and HPA design techniques due to the fact that the combination of both processing technology and circuit design techniques is essential for an optimum HPA performance. This work focuses on both topics. On the one hand, it studies the impact of different technology modifications on circuit designs with regard to the achievable efficiency with each technology. On the other hand, it deals with innovative HPA design techniques, mainly for X-band radar applications between 8 - 12 GHz, which have the potential to improve the overall performance with the main aim of a significant efficiency enhancement.

### 1.1.1 Advantages over existing III-V Technologies

Compound semiconductors play an important role in microwave power applications. Factors such as minimal weight, size, and power consumption are particularly important for aerospace self-sustaining systems. In addition, performance issues in X-band radar applications require the use of advanced components which enable high efficiency and high output-power levels to meet future demands.

The standard solution for X-band radar applications has classically been GaAs based, utilizing either MESFET or pHEMT device structures. However, the difficulty with these approaches has been the relative low voltage operation that restricted the device peak power handling [12].

From their basic material characteristic perspective, GaN-based electronic devices such as AlGaN/GaN high electron mobility transistors (HEMTs) are a good choice to overcome this problem. In addition, they promise even greater performance improvements in the future [84].

RF power electronics for high-power and high-frequency applications operated at high voltages imply the need for a wide bandgap material such as gallium nitride due to its outstanding material properties. The most important quality characteristics of Si, GaAs, 4H-SiC, and GaN are summarized and compared in Table 1.1.

On the one hand, the wide bandgap of GaN devices results in a high breakdown voltage that is about ten times higher than that

Table 1.1: Comparison of RF semiconductor material properties after [70].

| **Material** | **Si** | **GaAs** | **4H-SiC** | **GaN** |
|---|---|---|---|---|
| Relative dielectric constant $\varepsilon_r$ | 11.8 | 13.1 | 10 | 9 |
| Bandgap $E_g$ (eV) | 1.1 | 1.42 | 3.26 | 3.39 |
| Lattice constant (Å) | 5.4 | 5.7 | 3.1 | 3.2 |
| Electron mobility $\mu$ at 300 K ($cm^2/Vs$) | 1350 | 8500 | 700 | 1200-2000 |
| Saturated electron velocity $v_{sat}$ ($10^7 cm/s$) | 1 | 1 | 2 | 2.5 |
| Breakdown field $E_{br}$ (MV/cm) | 0.3 | 0.4 | 3 | 3.3 |
| Transit frequency $f_T$ (GHz) FET | 20 | 150 | 20 | 150 |
| Thermal conductivity $\Theta$ (W/cmK) | 1.5 | 0.43 | 3.3-4.5 | 1.3 |

reached with Si or GaAs. Therefore, operation at much higher voltages is possible which results in a reduced need for voltage conversion. As a consequence of the high voltage operation, GaN provides more than a factor of ten improvement in power density compared to GaAs devices, which results in higher output-power levels.

On the other hand, the requirement for high-frequency operation at least through X-band frequencies up to 12 GHz can be obtained with the wide bandgap materials GaN and SiC due to their high saturated electron velocities [70], [144] in combination with their high current density. The potential for operation at high power levels and at high frequencies makes GaN ideal for high-power marine, airborne and space-borne radar and communication applications.

Furthermore, devices fabricated from wide bandgap materials can be operated at much higher temperatures compared to their Si and other III-V counterparts [104], but with a significant lower drain current at higher temperatures due to electron satuation velocity reduction at high temperatures. Here, another aspect becomes important, the selection of the substrate. In general, GaN devices are fabricated on either SiC, sapphire, or Si substrates. The Fraunhofer Institute of Applied Solid-State Physics (IAF) technology uses SiC substrates due to its close lattive match to GaN and the superior thermal conductivity of >3 W/cm K leading to substantial higher output-power densities of GaN HEMTs compared to transistors made on, e.g., sapphire (see Table 1.1). Therefore, SiC is a good substrate candidate for power devices even though it is more expensive in fabrication than sapphire and Si. Thus, the high breakdown voltage and good thermal properties of the AlGaN/GaN material system on SiC substrates enable new levels of power amplifier performance.

In addition, the beneficial properties of an AlGaN/GaN HEMT lead to important advantages on design level compared to other existing III-V technologies, e.g. GaAs. For example, it is well known that the bandwidth of a high-power amplifier is limited by large input capacitances of large periphery devices capable to provide high power. Furthermore, the impedance transformation from a low optimum load impedance up to the 50 Ω environment also limits the bandwidth. To obtain a wide bandwidth and high power simultaneously, a device with low device capacitance and high optimum load impedance is required. Since GaN devices offer a high output-power density and good thermal properties, the fabrication of compacter devices with the same output power as their much larger GaAs counterparts is possible. The smaller size of the devices, in turn, leads to a reduction in terminal feedback capacitance and an increase of the output impedance for a given output power. Consequently, GaN devices enable an impedance matching over a wider bandwidth as well as a reduced matching effort and likewise lower net insertion losses.

The development of this relatively new semiconductor technology will further shift the boundary from vacuum tubes that currently dominate the high-power class at the higher frequency bands [64].

## 1.2 State-of-the-Art of X-Band MMICs

Over the last few years, the focus has strongly shifted to GaN-devices so that GaN development activities have dramatically increased. The transition from GaAs to GaN was a logical move considering the advances of GaN compared to GaAs in the field of high-efficiency and high-power amplifiers.
The trade-off between high-power and high-efficiency is one of the major challenges for the design of X-band HPA MMICs. Figure 1.1 represents the state-of-the-art power-added efficiency (PAE) results of HPA MMICs from the last five years operated at X-band frequencies as a function of the obtainable output power. The figure covers

state-of-the-art results of GaAs pHEMT based HPA MMICs [18], [47], [121], [30], [99] in form of black squares and GaN HEMT based power amplifiers. The GaN HPAs are divided into three categories. The first group includes results from different publications that are marked as blue crosses and that were found in [8], [23], [50], [71], [79], [80], [87], [100], [102], [115]. The other GaN MMICs marked as green circles and red stars were designed and developed at the Fraunhofer IAF, the red stars indicating particular HPA results from this work [88], [86], [55].

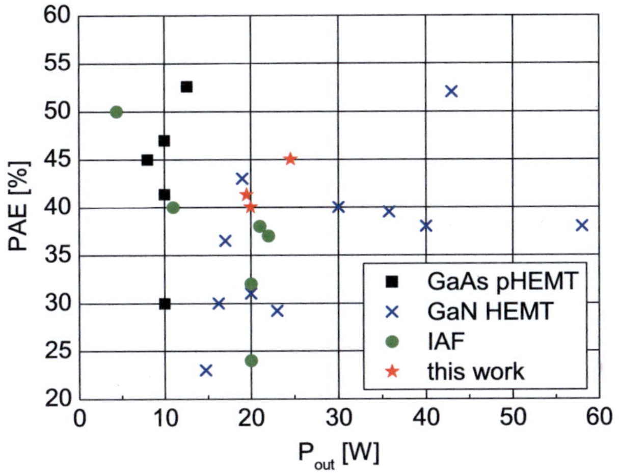


Figure 1.1: State-of-the-art PAE versus output power dependency of HPA MMICs from the last five years operated at X-band frequencies.

As can be seen in the figure, the GaAs pHEMT technology is limited to comparably low output-power levels of approximately 10 W. A maximum output power of 12.6 W could be obtained by Chu [16]. Though, typical X-band radar applications require output-power levels greater than 20 W on MMIC level, and in the near future even at module levels. According to the figure and from the general material properties of GaN, there are no reasons why GaN HEMT HPAs should not reach high output power levels above 20 W or even 30 W. Therefore, the focus has strongly shifted to GaN-based power amplifiers that are suited to provide high output-power levels.

However, the main problem of the relatively new GaN technology is the obtainable lower PAE that is considerably inferior compared to the PAE of GaAs-based HPAs. This fact is clearly shown in the figure. While three out of five GaAs-based publication examples offer a PAE beyond 45 %, only one GaN HEMT HPA outperforms the desired 45 % PAE boundary besides two Fraunhofer IAF HPA examples.

The Fraunhofer IAF technology based amplifiers show outstanding results at the front of the center span providing both a high output power of above 20 W and PAE values higher than 40 %. These results are only outperformed by a few single-frequency or narrow-band results reported in [79], [80]. In either case, the results of [79], [80] indicate that a further performance increase with GaN-based HPAs is possible.

However, there are different reasons for the inferior PAE behavior of GaN HEMTs compared to GaAs devices. First of all, the GaN-based transistor technology development is still in progress and has not yet reached maturity. As a result, the efficiency performance of GaN HEMTs is mainly limited due to device properties that still show e.g. higher DC-to-RF dispersion and higher drain leakage currents than their GaAs counterparts. The main influencing factors on the PAE performance of AlGaN/GaN HEMTs are shown in Chapter 2.3. In addition, the effective use of the high-impedance level of GaN devices for the design of high-power amplifiers needs to be further investigated.

This work has arisen out of the strong demand for a superior power-added efficiency of high-power amplifiers. Different concepts and approaches in order to obtain improved PAE values were analyzed. For the above-stated applications, it is obvious that every single percent in PAE enhancement delivers a strong positive impact on aerospace self-sustaining systems.

Therefore, the first part of this work includes the evaluation of the current Fraunhofer IAF GaN transistor performance (see Chapter 2) as well as the specific transistor technology progress and its impact on MMIC design level as described in Chapter 4. In the second part of this work, the implementation of well-known design measures from GaAs-based HPAs into GaN HEMT power amplifiers is analyzed in consideration of the electrical properties of GaN HEMTs such as the high impedance level of GaN devices. Three topics are discussed in detail:

1. Improved matching capabilities due to the design of a balanced amplifier taking advantage of the GaN-specific high load-impedance level (Chapter 5)

2. PAE enhancement in consideration of an optimized power- to driver-stage gate-width ratio (Chapter 6) and

3. Higher PAE performance by means of second harmonic termination (Chapter 6).

## 1.3 Outline

The scope of this thesis is the analysis as well as the design and layout of AlGaN/GaN high-power amplifier (HPA) MMICs for radar applications operating at X-band frequencies between 8 GHz and 12 GHz with a significant improvement of power-added efficiency (PAE). The evolving GaN HEMT technology provides enormous potential for improvements on the circuit level based on the higher impedance with nearly the same capacitances and transconductances $g_m$/mm compared to GaAs devices. The core areas that were examined in the course of this work are described in the following chapters.

Subsequent to the introduction, Chapter 2 focuses on the Fraunhofer IAF AlGaN/ GaN HEMT device including its principal properties and outstanding status of current RF-performance at X-band frequencies. Furthermore, the performance limitations, especially the PAE limiting factors, are defined and analyzed. In addition, their impact on the PAE behavior of the HEMT device is exemplarily demonstrated.

Chapter 3 gives an overview of the GaN-specific power-amplifier circuit design at X-band frequencies. The chapter covers all major design steps beginning with the transistor characterization at high-power operation under optimum load conditions, followed by the specific matching techniques for the different matching networks and small-signal and large-signal simulations as well as stability considerations. Finally, the circuit design cycle is completed with the layout aspect of the power amplifier.

Based on the HPA circuit-design expertise, Chapter 4 treats the development efforts regarding the AlGaN/GaN HEMT processing technology by means of real HPA MMICs processed with the different technologies. The aim of this chapter is the investigation of the impact of different technology modifications on circuit designs. Hereby, special attention is paid to the PAE performance of the HPA MMICs.

Chapter 5 covers the design of the first balanced microstrip HPA demonstrator in GaN technology for X-band taking advantage of the GaN-specific high load-impedance level. Since the major challenge in the design of a balanced HPA consists in the dimensioning and realization of the input and output couplers, the use of low-impedance microstrip input and output Lange couplers was investigated in this study to enable realizable dimensions of the couplers and to ensure a sufficient power and current handling capability. With the balanced HPA a strongly improved input and output matching capability of better than -14 dB could be obtained over the entire bandwidth from 8.5 - 13 GHz compared to -5 dB for the single-ended approach.

Furthermore, Chapter 6 features two efficient circuit design methods to improve the power-added efficiency within single- and dual-stage high-power AlGaN / GaN HEMT amplifiers for X-band frequencies. The first part focuses on the impact of the final (PA)- to driver (DRV)-stage gate-width (GW) ratio on the PAE performance of dual-stage HPAs. The study entails a detailed theoretical analysis and is supported by realized HPA MMICs offering different PA / DRV TGW ratios. In the second part, the principal effect of second harmonic termination at the input and/or output of a transistor on its PAE performance is analyzed in theory and proved on the basis of single-stage HPAs. In addition, unwanted side effects such as a limited bandwidth and a reduced output power performance are examined.

Finally, the work is summarized in Chapter 7 pointing out the main results of this thesis.

# Chapter 2

# High Electron Mobility Transistor

## 2.1 Principle Properties

The high electron mobility transistor (HEMT) is an advanced heterostructure device based on the MESFET principle [33]. However, due to its superior transport and general material properties (see Chapter 1), the HEMT device leads to higher gain, higher power-added efficiency (PAE), higher output power, and better low noise performance for a wider frequency range [83]. This makes the HEMT device most applicable for high-power applications.

**Small-Signal Model**

Figure 2.1 presents the small-signal model of a HEMT including the intrinsic device and parasitic elements. The figure gives an overview of the intrinsic and extrinsic parameters which are partially used with-in this work to explain their impact on the transistor performance.

The components $R_g$, $R_d$, and $R_s$ model the contact and metallization structure resistances and the elements $L_{g/d/s}$, $C_{pgi/pdi/pgdi}$ and $C_{pga/pda/pgda}$ indicate the extrinsic parasitic inductive and capacitive effects of the gate, drain and source, respectively [110].

The intrinsic small-signal model consists of eight components. $C_{gs}$ illustrates the charge modulation in the channel due to gate-source

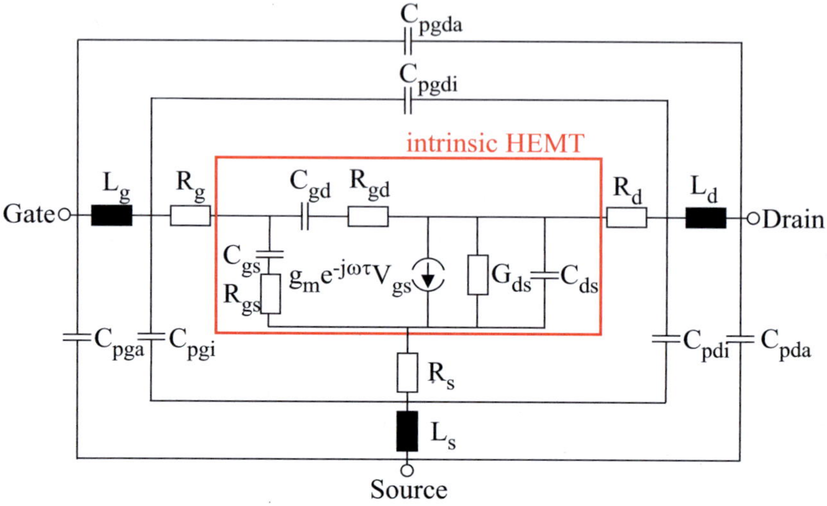


Figure 2.1: Small-signal equivalent circuit of a HEMT including parasitic elements.

voltage, $C_{gd}$ specifies the feedback capacitance, and $C_{ds}$ is a geometric capacitance between the drain and source that is mostly bias independent within the Fraunhofer Institute for Applied Solid State Physics (IAF) GaN HEMT technology. In addition, the device output conductance $G_{ds}$ is a measure of the incremental change in the output current $I_{DS}$ versus output voltage $V_{DS}$, while the input voltage $V_{GS}$ is held constant. $G_{ds}$ and $C_{ds}$ represent the output impedance.

The transconductance $g_m$ is defined as the ratio of $\Delta I_{DS}$ and $\Delta V_{GS}$ and the delay $\tau$ indicates the transit time associated with the carrier transport through the channel. Finally, $R_{gs}$ and $C_{gs}$ declare the gate-source impedance and $R_{gd}$ specifies the amount of feedback losses of the HEMT device. Y-parameter based calculations of the most important intrinsic parameters can be found in Appendix A.

**AlGaN/GaN HEMT Geometry and Technology Process**

In the course of this work, all AlGaN/GaN HEMTs are processed with the GaN25 process of the Fraunhofer IAF with a 0.25 $\mu$m gate length technology that is optimized for X-/Ku-band monolithic microwave integrated circuit (MMIC) applications from 6 to 18 GHz. In addition, the standard FET used within this work offers a gate geometry of 8x125 $\mu$m (number of gate-fingers x gate-width) with a gate-to-source spacing of 1 $\mu$m and a gate-to-drain spacing of 3.5 $\mu$m and with a gate-to-gate pitch $l_{gg}$ of 50 $\mu$m. The AlGaN/GaN heterostructures are grown on semi-insulating (s.i.) SiC substrates by Metal Organic Chemical Vapor Deposition (MOCVD).

Two different technologies, microstrip line (MSL) and coplanar waveguide (CPW), can be used for on-chip integration, i.e. the so-called monolithic integration of the entire circuit design including all passive components on one chip [48]. Both technologies consist of a substrate on which metal tracks are fabricated. The most important difference between the two technologies is related to the location of the ground plane. For microstrip technology it is located on the backside of the substrate, while coplanar technology has signal and ground metal on the topside of the substrate. This makes coplanar technology easier and cheaper to fabricate than microstrip. However, compared to microstrip, losses and dispersion of other modes such as the slotline mode are somewhat higher [44]. Another advantage of using MSL technology is its direct compatibility with commercial hybrid microstrip radar modules [75].

The processing of all HEMT devices and HPA MMICs within this work is performed in MSL technology consisting of frontside processing, substrate thinning to 100 $\mu$m, and backside processing including front-to-back substrate via holes [26]. Further specifications of the processing technique are given in [133].

## 2.2 DC- and RF-Performance

This section gives an overview of the current DC- and RF-performance of an 8x125 $\mu$m AlGaN/GaN HEMT at X-band frequencies by means of device measurement results.

The representative DC and pulsed-DC current-voltage (I-V) characteristics of the 1 mm device at 10 GHz are depicted in Figure 2.2. The DC output characteristics on the left side in Figure 2.2(a) were measured up to $V_{DS} = 60$ V with $V_{GS} = -4 \ldots 2$ V. A maximum drain-source current of approximately 840 mA/mm was obtained which is related to an aluminum mole fraction of 22 % for this technology (see Chapter 4.2.1).

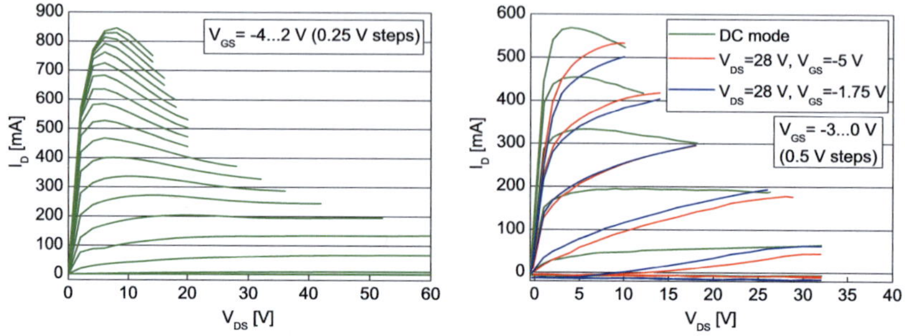


(a) DC I-V measurements in CW mode (b) DC and Pulsed I-V measurements

Figure 2.2: DC and Pulsed I-V measurements of an 8x125 $\mu$m AlGaN/GaN HEMT in microstrip technology with a gate length of 0.25 $\mu$m.

Since the RF GaN HEMT performance is affected by DC-to-RF dispersion effects and thermal phenomena, Figure 2.2(b) compares the DC (green lines) with the pulsed (red and blue lines) I-V curves in order to estimate the DC-to-RF dispersion. The DC-to-RF dispersion that is visible between the DC and pulsed-DC curves indicates the existence of trapping effects as explained in Section 2.3.3. As a result, $I_{\mathrm{D,\,max}}$ becomes reduced and $V_{\mathrm{k}}$ increased with a resulting degradation of the output power and a drop in the efficiency. Nevertheless, the DC-to-RF dispersion of the HEMT devices in the IAF GaN technology has been strongly improved within the last few years due to the HEMT-device technology progress. Temperature effects can be illustrated by comparing the two pulsed-DC I-V curves with the different quiescent bias points. The general shape of the curves remains unchanged, but the magnitudes decrease due to the higher channel temperature of the device measured with the higher gate quiescent biasing.

Next, Figure 2.3(a) shows the measured DC transconductance $g_{\mathrm{m}}$ of the same device sample. The transconductance shows an asymmetric behavior, i.e. a sudden rise near turn-on independent on $V_{\mathrm{DS}}$ followed by a smooth decrease toward $g_{\mathrm{m}} = 0\,\mathrm{mS}$ for $V_{\mathrm{GS}} > 1\,\mathrm{V}$ that becomes stronger for higher drain bias voltages. A maximum $g_{\mathrm{m}}$ of 310 mS/mm was measured at a drain bias voltage of 10 V with $V_{\mathrm{GS}}$ = -1.75... -1.5 V and 285 mS/mm at $V_{\mathrm{DS}}$ = 30 V with $V_{\mathrm{GS}}$ = -1.75 V

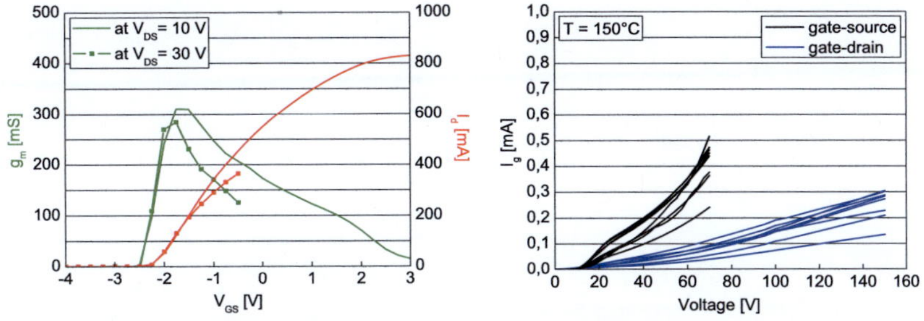


(a) Transconductance $g_m$ (b) Breakdown voltages $BV_{GS}$ and $BV_{GD}$

Figure 2.3: Transconductance and breakdown voltages of a 1 mm AlGaN / GaN HEMT in microstrip technology with a gate length of 0.25 $\mu$m measured at 10 GHz.

leading to correlating drain currents of each 100 mA. The complete channel pinch-off is at about -2.5 V independent of the drain bias voltage.

In addition, the gate-source and gate-drain breakdown behavior of the 0.25 $\mu$m HEMT were taken at an operation temperature of 150 °C. Mostly, the gate-drain breakdown voltage is above 100 V, which is about four times higher compared with GaAs devices having similar cut-off frequencies [81]. Figure 2.3(b) shows the breakdown behavior of the reverse biased gate-drain and the gate-source diode for ten 8x125 $\mu$m AlGaN / GaN HEMT devices from the same wafer as above. Both the gate-drain and the gate-source breakdown behavior were not measured up to breakdown. Nevertheless, the gate-source breakdown behavior of the majority of the measured transistors shows a strong increase in gate leakage current of up to 0.5 mA at 70 V. Therefore, the expected gate-source breakdown voltage causing a gate leakage current of 1 mA / mm is about 80 V. In addition, all devices offer an outstanding breakdown voltage for the gate-drain diode that is larger than 150 V.

The device also shows high cut-off frequencies that are extrapolated from the measured small-signal current gain $h_{21}$ and maximum stable gain (MSG) / maximum available gain (MAG) curves (see pages 26 and 40) as shown in Figure 2.4. An extrinsic transit frequency of $f_T = 30$ GHz and an extrinsic maximum frequency of oscillation of $f_{max} = 42$ GHz were obtained at $V_{DS} = 30$ V.

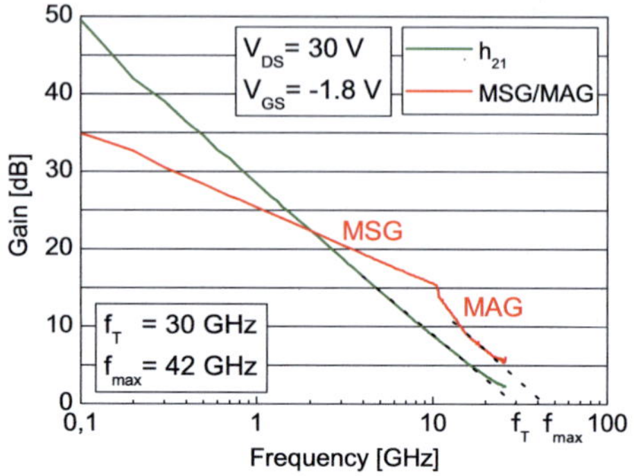


Figure 2.4: Measured current gain $h_{21}$ and small-signal power gain MSG/MAG of an 8x125 $\mu$m AlGaN/GaN HEMT and extrapolated $f_T$ and $f_{max}$. Bias point: $V_{DS}$= 30 V and $V_{GS}$= -1.8 V.

Figure 2.5 shows CW-load-pull results of a 8x125 $\mu$m AlGaN/GaN HEMT measured at 10 GHz and $V_{DS}$= 30 V in Class-AB (see page 30) operation without intentional harmonic matching. In Figure 2.5(a), the obtained $P_{out}$, PAE, and power gain are depicted as a function of $V_{DS}$. While the output power and power gain increase with increasing $V_{DS}$, the PAE performance remains constant at a value of nearly 45 % up to a drain voltage of 30 V. For $V_{DS}$> 30 V, the PAE performance

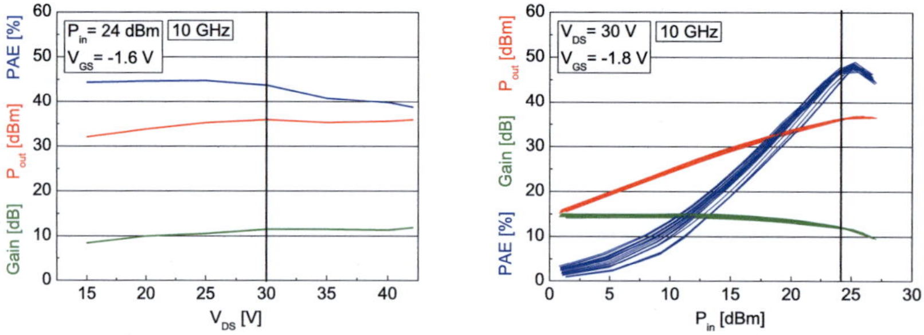


(a) $P_{out}$, PAE, and gain vs. $V_{DS}$ (b) Power-sweep across 21 cells vs. $P_{in}$

Figure 2.5: CW-load-pull measurements of a 1 mm AlGaN/GaN HEMT at $V_{DS}$= 30 V.

starts decreasing strongly due to the device characteristic and the fact that for high drain voltages the high magnitude of the optimum load impedance $\Gamma_L$ can not be set up any more due to limitations of the measurement setup. Nevertheless, at a drain voltage of 40 V, a PAE of 40 % can still be proven. Figure 2.5(b) shows a power-sweep of the same HEMT device across all 21 cells on a 3-inch wafer. As can be seen in the figure, the measurements offer both an excellent RF-performance and a very good on-wafer homogeneity, especially for the output power and gain performance.

A maximum PAE of 49 % was obtained at $P_{in} = 25$ dBm with an associated output power and power gain of 36.6 dBm and 11.3 dB, respectively. With a linear gain of 14.8 dB, the gain compression at the input power level of maximum obtainable PAE is 3.5 dB. In addition, a saturated output power density of 4.7 W/mm could be obtained.

Figure 2.6 shows the load-pull PAE measurement results as a function of the output power. In Figure 2.6(a) a mapping across all 21 cells of the entire wafer is depicted, measured for $P_{in} = 0 \ldots 27$ dBm with a fixed $\Gamma_L$. The figure on the right (Figure 2.6(b)) shows only one device sample measured at a fixed input power level and a variation of $\Gamma_L$. As can be seen especially in this figure, it is not possible to achieve a maximum PAE and maximum output power simultaneously. Depending on $\Gamma_L$, a maximum PAE of 49 % or a maximum output power

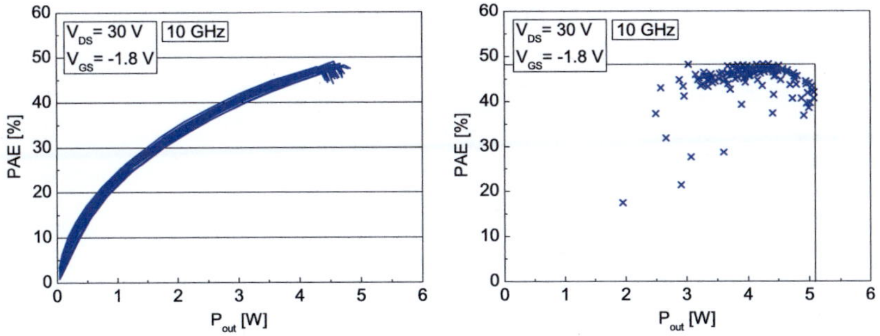


(a) Mapping results with fixed $\Gamma_L$

(b) $\Gamma_L$-Variation of one power cell

Figure 2.6: PAE vs. $P_{out}$ load-pull measurement results of an 8x125 $\mu$m AlGaN/GaN HEMT at 10 GHz in CW-mode. Bias point: $V_{DS}$=30 V and $V_{GS}$=-1.8 V.

greater than 5 W could be obtained.

Finally, Table 2.1 summarizes the most important DC and RF parameters of the 8x125 $\mu$m $Al_{0.22}Ga_{0.78}N$/GaN HEMT with $l_g$=0.25 $\mu$m.

According to the excellent results that are shown in the figures and listed in Table 2.1, the GaN25 Fraunhofer IAF process is a reliable and highly competitive technology. Furthermore, it is an excellent basis for further performance enhancement, especially regarding the PAE.

Table 2.1: Measured DC and RF parameters of a 1 mm AlGaN/GaN HEMT with $l_g$ = 0.25 $\mu$m.

| **DC parameters** | **Value** | **RF parameters measured at $V_{DS}$= 30 V and 10 GHz** | **Value** |
|---|---|---|---|
| $I_{D,max}$ | 840 mA/mm | $P_{out}$ ($P_{in}$ = 25 dBm) | 36.6 dBm |
| $g_{m,10V}$ | 310 mS/mm | PAE ($P_{in}$ = 25 dBm) | 49 % |
| $g_{m,30V}$ | 285 mS/mm | $G_p$ ($P_{in}$ = 25 dBm) | 11.3 dB |
| $BV_{GS}$ (150 °C) | >70 V | $G_{lin}$ | 14.8 dB |
| $BV_{GD}$ (150 °C) | >150 V | $P_{sat}$ | 4.7 W/mm |
| $f_T$ | 30 GHz | $P_{max}$ ($\Gamma_{L,Pout}$) | 5.1 W/mm |
| $f_{max}$ | 42 GHz | | |

## 2.3 Influencing Factors on PAE Performance

The power and efficiency performance of an AlGaN/GaN HEMT is mainly limited due to the following factors (three static and three dynamic phenomena):

- Knee voltage $V_k$
- Drain leakage current $I_{min}$
- DC-to-RF dispersion $\Delta I_{DC\text{-}RF}$
- Feedback
- Compression behavior
- Thermal effects and reliability.

Figure 2.7 shows the impact of the first three influencing factors by means of the I-V characteristic of an AlGaN/GaN HEMT.

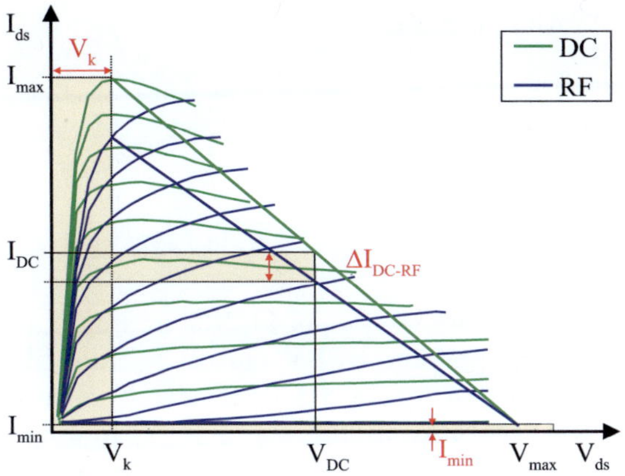


Figure 2.7: Static PAE influencing factors marked in red in the I-V characteristic of an AlGaN/GaN HEMT.

## 2.3.1 Knee Voltage

The knee voltage is defined as the voltage of the I-V curves transition from "linear" to "saturation". Here, the knee turn-on region has a substantial impact on the HEMT performance, especially the results for output power and efficiency due to the fact that $V_k > 0$ causes an increase in the DC supply power. The DC supply power is the well-known product of the DC supply voltage and current.

$$P_{DC} = V_{DC} \cdot I_{DC} \qquad \text{with} \quad V_{DC,real} = V_{DC,ideal} + V_k \tag{2.1}$$

Besides, the efficiency of a transistor can be defined in different ways as shown in Chapter 3.1.2. Within this work, the so-called power-added efficiency (PAE) of a transistor is used, since the PAE indicates the transistor's efficiency under consideration of the required input power level $P_{in}$ which is substantial for a HPA design. The PAE is defined by

$$PAE = \frac{P_{out} - P_{in}}{P_{DC}} = PAE_{ideal} \,. \tag{2.2}$$

To obtain the impact of the knee voltage on the PAE of a transistor, the above stated equations are combined to the following more precise ("real") PAE equation:

$$\begin{aligned} PAE_{\text{real}} &= \frac{P_{\text{out}} - P_{\text{in}}}{(V_{\text{DC,ideal}} + V_{\text{k}}) \cdot I_{\text{DC}}} \\ &= PAE_{\text{ideal}} \frac{1}{1 + V_{\text{k}} / V_{\text{DC, ideal}}} . \end{aligned} \tag{2.3}$$

For a typical 1 mm AlGaN/GaN HEMT operated at X-band frequencies with a knee voltage of 5 V and at a drain bias voltage of $V_{\text{DC}} = 35$ V, the impact of $V_{\text{k}}$ on the PAE performance is a factor of about 0.875 of the ideal PAE assuming $V_{\text{k}} = 0$. In other words, e.g., an ideal PAE of 70 % would be reduced down to 61.25 %. As can be seen in the example, the knee voltage has a deep impact on the PAE behavior of a transistor and therefore on the HPA design due to the fact that it will always be a significant percentage of the DC supply [25].

The walk-out of the knee voltage arises from the lengthening of the gate-drain space-charge region that is controlled by the bias point [129]. Therefore, the RF knee voltage increases with increasing drain bias voltages that cause a significant rise in $R_{\text{on}}$. As a result and based on the fact that $V_{\text{k}}$ is proportional to the product of $R_{\text{on}}$ and the saturated drain current [28], the obtainable PAE is decreasing towards higher drain biases [43].

A reduction in knee-voltage can be obtained for field-plated devices [14] as discussed in detail in Chapter 4. The field-plate leads to a reduced DC-to-RF dispersion so that the RF-knee voltage becomes smaller and aspires towards the DC knee voltage.

### 2.3.2 Drain Leakage Current

According to the knee voltage, the drain leakage current $I_{\text{min}}$ also affects the PAE performance of a HEMT device. It causes reliability problems [27], and results in a lowered breakdown voltage and an increased OFF-state loss [98], [63]. The relationship between the drain leakage current and the impact on the PAE performance can be illustrated using an equivalent formula as for the PAE related to the knee voltage, i.e. Equation 2.3.

$$PAE_{\text{real}} = \frac{P_{\text{out}} - P_{\text{in}}}{V_{\text{DC}} \cdot (I_{\text{DC,ideal}} + I_{\text{min}})} = PAE_{\text{ideal}} \frac{1}{1 + I_{\text{min}} / I_{\text{DC}}} \,. \tag{2.4}$$

Again, taking a realistic drain leakage current of $I_{\text{min}} = 1\,\text{mA}$ into account within a Class-AB to Class-B operation with $I_{\text{DC}} = 1/\pi\,\text{A}$ (see Section 3.1.2, Table 3.1), the real PAE is approximately a factor of 0.99 of the ideal PAE. Thus, the drain leakage current in the order of 1 mA delivers almost no impact on the PAE performance compared to a knee voltage of 5 V.

### 2.3.3 DC-to-RF Dispersion

Another $P_{\text{out}}$ and PAE degradation factor for power FETs is the discrepancy between the expected output power from the DC current-voltage characteristic of the FET compared with the compressed RF current swing leading to reduced load-pull measurement results of $P_{\text{out}}$ [18]. The resulting difference between the DC and RF loadlines is shown in Figure 2.7. The DC-to-RF discrepancy is attributed to the trapping effects where both surface and bulk traps contribute [143], [9], [128]. Hereby, the surface trapping effects significantly impact the problem of drain current collapse [129].

The so-called DC-to-RF dispersion affects the device performance due to a reduction in saturation current and an increase in knee voltage at high frequencies and high bias conditions [20]. Its impact on the PAE performance can be calculated in accordance with the last two PAE equations and with respect to $\Delta I_{\text{DC-RF}} = (I_{\text{max}} - I_{\text{min}})_{\text{DC}} - (I_{\text{max}} - I_{\text{min}})_{\text{RF}}$.

$$PAE_{\text{real}} = \frac{P_{\text{out}} - P_{\text{in}}}{V_{\text{DC}} \cdot (I_{\text{DC,ideal}} + \Delta I_{\text{DC-RF}})} = PAE_{\text{ideal}} \frac{1}{1 + \Delta I_{\text{DC-RF}} / I_{\text{DC}}} \,. \tag{2.5}$$

Assuming that the discrepancy between the DC and RF I-V characteristics of a 1 mm AlGaN/GaN HEMT operated at X-band frequencies amounts $\Delta I_{\text{DC-RF}} / I_{\text{DC}} = 0.1$, the resulting real PAE yields 90 % of the ideal case without DC-to-RF dispersion.

To mitigate the DC-to-RF dispersion, surface SiN passivation was optimized to stabilize surface states and therefore recover surface trapping effects that are responsible for limiting both the RF current and breakdown voltages of the device [42], [67]. In addition, the adoption of field modulating plates (FP) also improves the trade-off relation between current collapse and breakdown characteristics and therefore leads to a reduced dispersion and a resulting higher output power and PAE [3], [140], [70]. The impact of field plates on the performance of HPA designs is demonstrated in Chapter 4. Nevertheless, the field plate causes additional gate capacitance that in turn leads to a degraded high-frequency performance of the power HEMT device for millimeter-wave amplifiers. Therefore, in case of high-frequency applications, a deeply recessed gate structure and an adequate variation of the GaN cap thickness [134], [131] that both have a strong impact on the DC-to-RF dispersion are preferred to the use of field plates [18], [108]. Furthermore, a V-shaped gate also delivers the desired reduction in DC-to-RF dispersion and therefore leads to a performance improvement [17].

### 2.3.4 Feedback

Besides the above stated obvious PAE degradation factors, the feedback effect should also not be neglected when the HEMT is operated at high voltages due to the voltage dependency of the feedback capacitance $C_{\text{gd}}$. In addition, the high-frequency performance is limited by $C_{\text{gd}}$ and the feedback resistance $R_{\text{gd}}$. On the one hand, $R_{\text{gd}}$ can be reduced by the use of an appropriate gate geometry as discussed in Chapter 4. On the other hand, the feedback capacitance that is mainly located between the top of the gate and the GaN cap layer can be reduced by an adapted SiN passivation layer [103] and source-connected field plates as demonstrated in Chapter 4. However, the use of source-connected field plates causes a strong increase in $R_{\text{gd}}$ as is documented in Chapter 4.1.3.

Figure 2.8 shows the small-signal equivalent circuit of a HEMT device with the feedback admittance $Y_{\text{FB}}$. This circuit can be transformed by means of a compensation of $(C_{\text{ds}} + C_{\text{gd}})$ with $L_{\text{comp}}$ that is described in detail in Chapter 3.1.4. The resulting circuit indicates

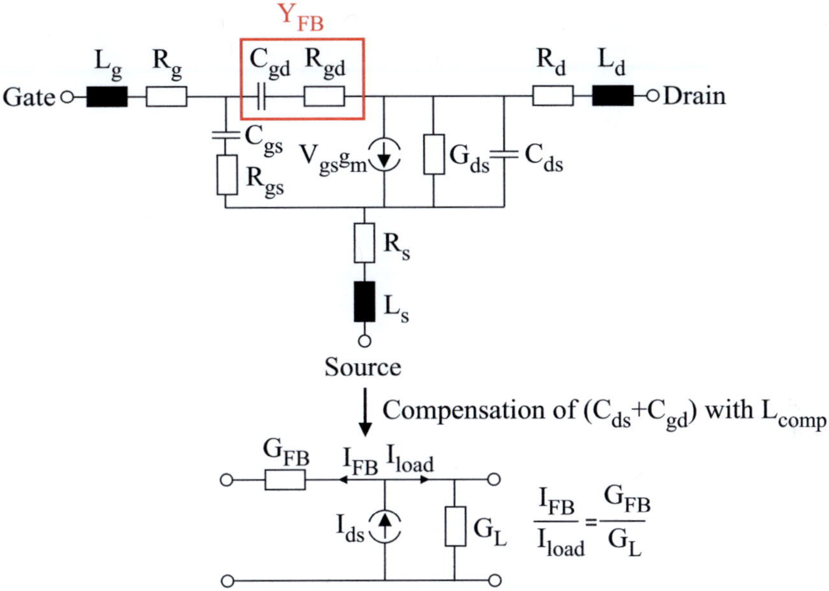


Figure 2.8: Small-signal equivalent circuit of a HEMT device including the feedback admittance $Y_{\mathrm{FB}}$.

the relation between the ratio of feedback and load currents and the ratio of feedback and load conductances, i.e. $I_{\mathrm{FB}}/I_{\mathrm{load}} = G_{\mathrm{FB}}/G_{\mathrm{L}}$. Hereby, the feedback admittance can be calculated as

$$Y_{\mathrm{FB}}(\omega) = \frac{1}{R_{\mathrm{gd}}(\omega) + \frac{1}{j\omega C_{\mathrm{gd}}(\omega)}} = G_{\mathrm{FB}}(\omega) + j\omega C_{\mathrm{FB}}(\omega)\ . \tag{2.6}$$

According to Figure 2.8, the ratio of the feedback power and the output power amounts

$$\frac{P_{\mathrm{FB}}}{P_{\mathrm{out}}} = \frac{1/2\,|I_{\mathrm{FB}}|^2\,1/G_{\mathrm{FB}}}{1/2\,|I_{\mathrm{load}}|^2\,R_{\mathrm{L}}} = G_{\mathrm{FB}} R_{\mathrm{L}}\ . \tag{2.7}$$

According to [122], the impact of the feedback effect on the PAE performance of a HEMT device results with the above stated power relation in

$$PAE_{\mathrm{real}} = PAE_{\mathrm{ideal}} \cdot (1 - \frac{P_{\mathrm{FB}}}{P_{\mathrm{out}}})$$
$$= PAE_{\mathrm{ideal}} \cdot (1 - \omega^2 C_{\mathrm{gd}}^2 R_{\mathrm{gd}} R_{\mathrm{L}}) . \quad (2.8)$$

With typical values of $C_{\mathrm{gd}} = 0.1\,\mathrm{pF}$, $R_{\mathrm{gd}} = 20\,\Omega$, and $R_{\mathrm{L}} = 60\,\Omega$ of a 1 mm AlGaN/ GaN HEMT operated at 10 GHz, the real PAE would yield 95 % of the ideal PAE.

### 2.3.5 Compression Behavior

Another important aspect in terms of the obtainable PAE performance is the gain compression behavior of the HEMT that is caused by the non-linear characteristics of the HEMT device when operated at high input power levels and defined as the difference between the linear and power gain. The gain compression is mainly dependent on the $g_{\mathrm{m}}$ curve progression that is shown in Figure 2.3(a) for a typical AlGaN/GaN HEMT in Fraunhofer IAF technology. The curve characteristic offers an asymmetric behavior with a sudden rise near turn-on. This strong increase in $g_{\mathrm{m}}$ enables the operation at $g_{\mathrm{m,\,max}}$ and at a simultaneous small drain current of about one-tenth of $I_{\mathrm{D,\,max}}$. However, the slow decrease towards $g_{\mathrm{m}} = 0\,\mathrm{mS}$ causes already a non-linear behavior of $I_{\mathrm{D}}$ and therefore a decrease in gain. To obtain saturated output power, a gain compression of at least 3 dB is necessary for AlGaN/GaN HEMT devices in the Fraunhofer IAF technology.

According to Equation 2.2, the PAE is dependent on the relation of the output and input power level, i.e. the gain performance. Therefore, Equation 2.2 can be restated as

$$PAE = \frac{P_{\mathrm{out}}}{P_{\mathrm{DC}}}(1 - \frac{1}{G}) . \quad (2.9)$$

As can be seen in Equation 2.9, the PAE increases with increasing gain performance. With a characteristic power gain of 10 dB the PAE amounts 90 % of the relation of $P_{\mathrm{out}} / P_{\mathrm{DC}}$ which is defined as the drain efficiency $\eta$ of the device.

### 2.3.6 Thermal Effects and Reliability

GaN based semiconductor devices on SiC substrate offer a high thermal conductivity that provides a great potential for high temperature application fields. However, the "operating" temperature has a significant impact on both the performance and reliability of the device. As the active region (channel) temperature increases, the crucial device parameters such as the maximum operating frequency, the electron saturation velocity, the output power, gain, and accordingly the PAE performance of the HEMT device decrease [10], [35]. Furthermore, it is commonly assumed that the reliability of a transistor also decreases with rising temperature.

Taking all the above-mentioned examples from 2.3.1 - 2.3.5 together, the realizable power-added efficiency amounts only:

$$\begin{aligned} PAE &= \eta \cdot \frac{1}{1+V_{\mathrm{k}}/V_{\mathrm{DC}}} \frac{1}{1+I_{\mathrm{min}}/I_{\mathrm{DC}}} \frac{1}{1+\Delta I_{\mathrm{DC\text{-}RF}}/I_{\mathrm{DC}}} \cdot \\ &\quad \cdot (1-\omega^2 C_{\mathrm{gd}}^2 R_{\mathrm{gd}} R_{\mathrm{L}})(1-\frac{1}{G}) \\ &= \eta \cdot 0.875 \cdot 0.99 \cdot 0.9 \cdot 0.95 \cdot 0.9 = \eta \cdot 0.67 \end{aligned} \tag{2.10}$$

Thus, these PAE influencing factors have a strong impact on the PAE performance. To achieve PAE values above the calculated 67 % of the drain efficiency from the single examples, the AlGaN/GaN HEMT performance needs to be improved by an advanced processing progress. Within the last few years, the PAE could be strongly improved by means of the above mentioned technology enhancements such as the introduction of field modulating plates. An overview of different technology improvements related to a PAE performance optimization is given in Chapter 4 on the basis of high-power amplifiers processed with the continuously improving technology.

# Chapter 3

# Power-Amplifier Circuit Design

In this chapter, the design of an AlGaN/GaN high-power amplifier (HPA) with its fundamental aspects will be explained. Figure 3.1 gives a general overview of the design flow of a typical power amplifier including all major design steps that are discussed in this chapter. Furthermore, GaN-specific factors such as the high impedance level of AlGaN/GaN transistors and the consequences for the HPA design are elaborated in this chapter.

**HPA Structure**

Initially, the selection of the amplifier's structure needs to be made. The available technology and the required amount of output power and power gain determine the number of amplifier stages and the number of transistors that must be used in parallel, especially in the output stage. The technology defines factors such as the transit frequency, the maximum supply voltage and the noise figure that influence the estimated gain, maximum output power and noise performance of an amplifier at a certain frequency. First of all, the specifications of the high-power amplifier have to be defined including the frequency-range, the drain bias voltage $V_{DS}$, the output power $P_{out}$, the power-added efficiency (PAE), and the associated power gain performance.

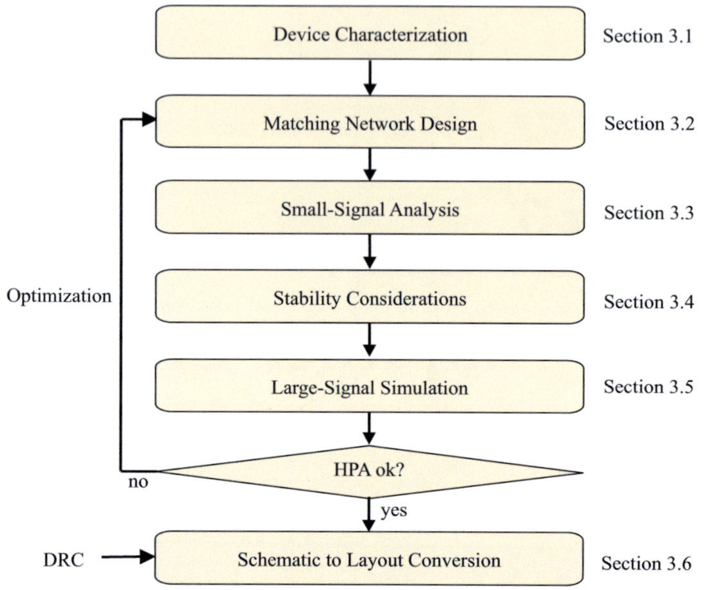


Figure 3.1: Design flow of a power amplifier including all major design steps.

The required total gate width of the output (power)-stage transistors can be calculated with respect to the required output power of the entire power amplifier, the expected loss of the output matching network (OMN), and the power density of the output-stage unit transistor cell. In the case of a dual-stage amplifier, the size and number of the FET cells in the driver-stage can be determined according to the transistor periphery of the PA-stage. In this case, the total gate width of the driver (DRV) has to be large enough to prevent the DRV-stage FET cells from going into compression before the power (PA)-stage transistors. The influence of the amplifier's structure, i.e. the gate width and gate-width ratio of both transistor stages, on the performance of the entire high-power amplifier will be discussed in detail in Section 6.1.

All HPAs in this work are required to operate within X-band frequencies between 8-12 GHz and at drain bias voltages above 30 V. Besides the "monitor" amplifiers (see Chapter 6.2), an output power beyond 20 W with an associated PAE $\geq 40\,\%$ is required for typical radar applications. In this work, only one- and two-stage amplifiers are discussed.

## 3.1 Device Characterization

For the design of high-power amplifiers, transistors with sufficiently high frequency and power performance are required. The layout of the unit transistor cell needs to be extracted from the estimated total gate width of the output-stage transistors and the correlating cell size of the single transistor. Geometry issues such as the gate width and the number of gate fingers have to be decided upon. In addition, the gate-to-gate pitch has an important effect on the layout and the characteristics of a transistor as discussed in the last chapter. While a reduction of this spacing results in a smaller device layout, it also leads to an increase of the maximum gate finger temperature with a resulting decrease of the transistor's reliability.

### 3.1.1 Transistor Size Selection

Initially, the relationship between the gate width and the number of gate fingers is analyzed on the basis of AlGaN/GaN HEMTs with a gate length of 0.25 $\mu$m and approximately 1 mm total gate width each; namely $8\times125\,\mu$m, $10\times100\,\mu$m, $12\times85\,\mu$m, and $16\times60\,\mu$m transistors in microstrip line (MSL) technology. Furthermore, all transistors contain source finger air bridge connections with four external source vias to ground, but no individual source vias (ISV). The layout of the $8\times125\,\mu$m HEMT cell is shown in Figure 3.2.

Figure 3.3 depicts a comparison of the maximum stable gain (MSG) and maximum available gain (MAG) curves of the above defined FET variants that are all biased at the same drain voltage of 30 V. The green area in both graphics presents the relevant X-band frequencies. The figure on the left shows the curve characteristics over a wide frequency-range up to 20 GHz in logarithmic scale, while the right figure zooms into the relevant X-band area as shown in the black rectangle from the figure on the left. As can be seen in the figures, the device with the minimum number of fingers delivers the highest MSG/MAG values, i.e. an improvement of 0.5-0.8 dB over the entire

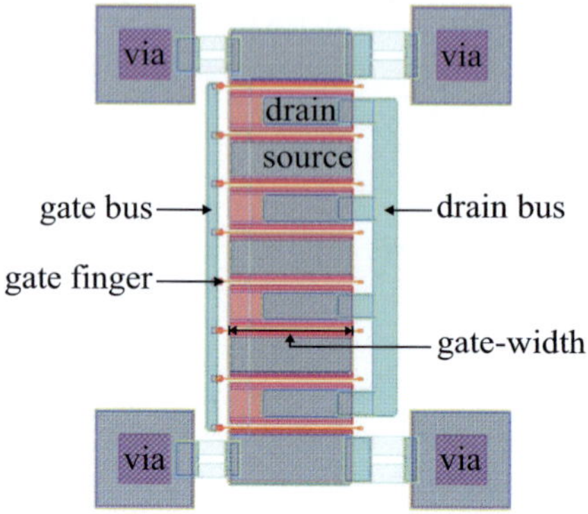


Figure 3.2: Typical pattern layout of an 8×125 $\mu$m HEMT cell with key dimensions.

X-band frequency-range compared with the device with 16 gate fingers. In addition, the k-point, which is defined by the frequency at which the Rollett-factor turns one (see pages 39/40), of the device with 8 gate fingers is 1 GHz higher than that of the device with 16 gate fingers. Nevertheless, the k-points of all devices are beyond the upper end of the X-band frequency range, i.e. 12 GHz.

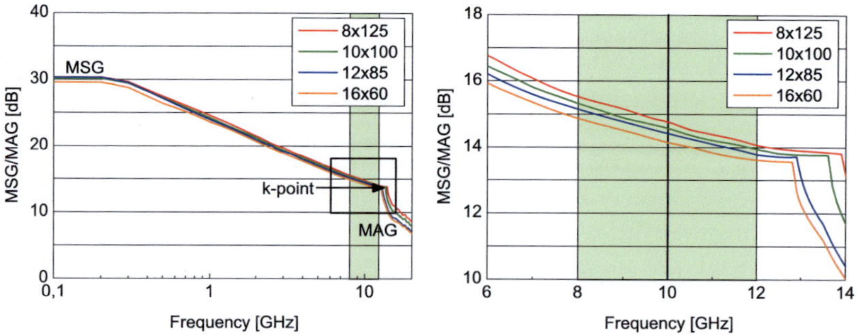


Figure 3.3: Comparison of the MSG/MAG characteristics of AlGaN/GaN HEMT variants each with approximately 1 mm gate geometry, but different gate widths and number of gate fingers. Bias point: $V_{DS}$= 30 V and $V_{GS}$= -2 V.

The higher MSG / MAG values of the devices with less finger count can be explained with help of Equations 3.1 and 3.2 according to [33] and [78], respectively.

$$\mathrm{MSG} = \left|\frac{S_{21}}{S_{12}}\right| = \left|\frac{Y_{21}}{Y_{12}}\right| \approx \frac{g_\mathrm{m}}{\omega C_\mathrm{gd}} \quad \text{(low-frequency approximation)} \tag{3.1}$$

$$\mathrm{MAG} \approx \left(\frac{f_\mathrm{T}}{f}\right)^2 \frac{1}{4\left[G_\mathrm{ds}(R_\mathrm{in} + \pi f_\mathrm{T} L_\mathrm{s}) + \pi f_\mathrm{T} C_\mathrm{gd}(R_\mathrm{in} + R_\mathrm{g} + 2\pi f_\mathrm{T} L_\mathrm{s})\right]} \tag{3.2}$$

On the one hand, the external capacitance $C_{gd}$ increases with increasing number of gate fingers due to the larger gate bus length, so that the maximum stable gain decreases. On the other hand, at X-band frequencies the layout of the HEMT device, in particular the extrinsic parasitic elements, influences the MAG behavior as can be seen in Equation 3.2. Again, with increasing finger count and consequent increasing source bus length, the source inductance $L_s$ also becomes larger leading to a reduction in the MAG behavior.

Thus, the parasitic elements caused by the supply paths and connections are mainly influenced by the layout structure of the HEMT device. In this study, the HEMT structure is analyzed only within one "direction", i.e. towards a higher number of gate fingers, based on the $8\times125\,\mu$m HEMT cell from Figure 3.2 with a nearly quadratic layout structure. The MSG / MAG behavior of HEMT devices with fewer finger counts than eight and consequently wider gate widths would result in comparable layout structures and therefore also lead to an inferior MSG / MAG performance. The "quadratic" layout structure delivers the highest MSG / MAG values. As a result, the transistor with the $8\times125\,\mu$m gate geometry is found to be the best device selection concerning the MSG / MAG performance. Thus, this geometry is extensively used for the designs within this work.

### 3.1.2 Class of Operation

The bias condition of an amplifier mainly depends on the amplifier requirements for output power, efficiency, linearity, and possible bandwidth. With the high-power amplifiers presented in this work, on the one hand, a preferably high output power is basically sought.

This leads to a Class-A or Class-AB operation. On the other hand, high efficiencies are required that can be realized with Class-B or Class-C configurations as shown in Table 3.1. Table 3.1 compares the most relevant amplifier specifications of this work, i.e. the drain efficiency $\eta_D$ and the realizable bandwidth $BW$, operating at different classical current source amplifier classes with an associated current conduction angle $\alpha$. Further characteristics of the amplifier's operating classes are listed in [33], [5].

Table 3.1: Overview of the commonly used "linear" operating classes of classical current-mode PAs.

| **Biasing** | $\eta_{D,max}$ | $\alpha$ | **Bias Point** | **Possible BW** |
|---|---|---|---|---|
| Class-A | 50 % | $2\pi$ | $I_{DC} = I_{max}/2$<br>$V_{DC} = (V_{max} + V_k)/2$ | high |
| Class-AB | 50 % - 78.5 % | $\pi$ - $2\pi$ | between A and B | high |
| Class-B | 78.5 % | $\pi$ | $I_{DC} = I_{max}/\pi$<br>$V_{DC} = (V_{max} + V_k)/2$ | moderate |
| Class-C | 78.5 % - 100 % | 0 - $\pi$ | $I_{DC} < I_{max}/\pi$<br>$V_{DC} = (V_{max} + V_k)/2$ | low |

The output power and efficiency of a transistor are directly influenced by the operating class of the transistor. Here, the efficiency can be measured as the drain efficiency which is determined as

$$\eta_D = \frac{P_{out}}{P_{DC}} \tag{3.3}$$

or as the power-added efficiency (PAE) which is defined as

$$PAE = \frac{P_{out} - P_{in}}{P_{DC}} = \frac{P_{out}}{P_{DC}} \left(1 - \frac{1}{G}\right) . \tag{3.4}$$

The use of the PAE is more useful because of the influence of the gain within the calculation [29]. For $G \to \infty$, the power-added efficiency is equal to the drain efficiency. The DC supply is given by $P_{DC} = V_{DC} \cdot I_{DC}$ whereas the corresponding current can be calculated by

Fourier analysis according to Cripps [25] yielding

$$
\begin{aligned}
I_{\mathrm{DC}} &= \frac{1}{2\pi}\int_{-\alpha/2}^{\alpha/2} I_d(\theta)\,d\theta \\
&= \frac{1}{2\pi}\int_{-\alpha/2}^{\alpha/2} \frac{I_{\mathrm{max}}}{1-cos\,(\alpha/2)}\cdot(cos\,\theta - cos\,(\alpha/2))\,d\theta \\
&= \frac{I_{\mathrm{max}}}{2\pi}\cdot\frac{2\cdot sin\,(\alpha/2)-\alpha\cdot cos\,(\alpha/2)}{1-cos\,(\alpha/2)}\,. \qquad (3.5)
\end{aligned}
$$

Figure 3.4 gives an overview of the bias point locations of a typical AlGaN/GaN-HEMT for all relevant classes of high-power amplifier operation in this work. On the left side, the drain current versus drain-source and gate-source voltage curves (I-V characteristics) and the optimum load line for a Class-A amplifier are shown. In addition, the figure on the right depicts the drain current versus gate-source voltage in saturation region. The corresponding bias point locations are marked in both figures.

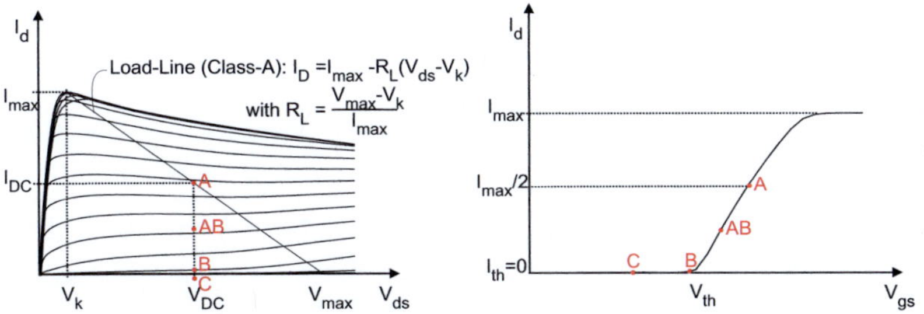


Figure 3.4: Bias point locations of a typical AlGaN/GaN-HEMT for various operation classes.

For the calculation of the maximum obtainable $\eta_{\mathrm{D}}$ in different operation classes, Figure 3.5 clarifies the loadlines of a Class-A and a single-ended and push-pull ("complementary amplifier" using reverse polarity transistors in Class-B operation to provide the full waveform) Class-B operation.

The corresponding DC supply power result is given in Equation 3.6 for Class-A operation, in Equation 3.7 for single-ended Class-B operation, and in Equation 3.8 for push-pull Class-B operation.

$$P_{\text{DC, A}} = \frac{V_{\max}+V_{\text{k}}}{2}\cdot\frac{I_{\max}}{2} = \frac{1}{4}\cdot V_{\max}I_{\max} \quad \text{with } V_{\text{k}}=0 \tag{3.6}$$

$$P_{\text{DC, B}}(\text{single-ended}) = \left(V_{\max}-\frac{V_{\max}-V_{\text{k}}}{\pi}\right)\cdot\frac{I_{\max}}{\pi} = \frac{\pi-1}{\pi^2}\cdot V_{\max}I_{\max} \quad \text{with } V_{\text{k}}=0 \tag{3.7}$$

$$P_{\text{DC, B}}(\text{push-pull}) = \frac{V_{\max}+V_{\text{k}}}{2}\cdot\frac{I_{\max}}{\pi} = \frac{1}{2\pi}\cdot V_{\max}I_{\max} \quad \text{with } V_{\text{k}}=0 \tag{3.8}$$

The RF power of a Class-A and a Class-B operation is defined as

$$P_{\text{RF, A}} = P_{\text{RF, B}} = \frac{V_{\text{RF}}}{\sqrt{2}}\frac{I_{\text{RF}}}{\sqrt{2}} = \frac{1}{2}\cdot\frac{V_{\max}-V_{\text{k}}}{2}\frac{I_{\max}}{2} = \frac{1}{8}\cdot V_{\max}I_{\max} \quad \text{with } V_{\text{k}}=0\,. \tag{3.9}$$

Consequently, the drain efficiencies for the above stated operation classes result in

$$\eta_{\text{D, Class-A}} = \frac{P_{\text{RF, A}}}{P_{\text{DC, A}}} = \frac{1}{2} = 50\,\% \tag{3.10}$$

$$\eta_{\text{D, Class-B}}(\text{single-ended}) = \frac{P_{\text{RF, B}}}{P_{\text{DC, B}}(\text{single-ended})} = \frac{\pi^2}{8(\pi-1)} = 57.6\,\% \tag{3.11}$$

$$\eta_{\text{D, Class-B}}(\text{push-pull}) = \frac{P_{\text{RF, B}}}{P_{\text{DC, B}}(\text{push-pull})} = \frac{2\pi}{8} = 78.5\,\%\,. \tag{3.12}$$

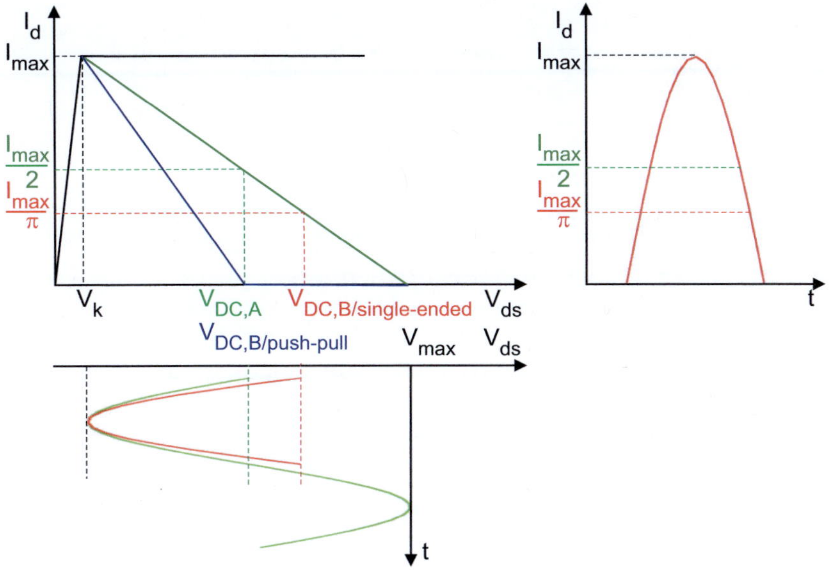


Figure 3.5: Voltage-current waveforms for a Class-A and a single-ended and push-pull Class-B operation.

**Class-A**

The DC bias point of a Class-A amplifier is the exact mid-point of the linear range as shown in Figure 3.4. Consequently, the signal is amplified during the full signal period ($\alpha = 2\pi$) without signal clipping, i.e. only the fundamental signal component exists, leading to a perfectly linear operation. Being operated at Class-A condition, the DC voltage is $(V_{\max} + V_{\mathrm{k}})/2$. From Equation 3.5, the corresponding DC current $I_{\mathrm{DC}}$ results in $I_{\max}/2$. The DC power of a Class-A amplifier is independent of the RF input or output power levels, which is one of the major drawbacks of Class-A amplifiers because they consume this amount of power even with no signal applied [130]. This leads to a poor maximum efficiency of $\eta_{\mathrm{D}}\,(\text{Class-A}) = 50\,\%$.

Class-A amplifiers are most commonly used in small-signal applications where linearity is more important than power efficiency. Nevertheless, Class-A amplifiers are also used in large-signal applications with an extraordinarily need for high linearity and high gain that outweighs the disadvantage of poor power efficiency [90].

**Class-B**

In Class-B operation, the quiescent DC bias current is set to zero so that the conduction angle of each RF cycle results in $\alpha = \pi$ (half sine waveform). That causes the RF-output-power to DC-source-power efficiency to be much higher than with Class-A amplifiers. Class-B delivers the same RF output power as Class-A (see Equation 3.9), but with a reduced DC current of $I_{DC}$ (Class-B) $= I_{max}/\pi$ according to Equation 3.5. Consequently, the DC supply is reduced by a factor of up to $2/\pi$ for a push-pull configuration compared with Class-A condition. This results in a theoretical maximum efficiency of $\eta_D$ (Class-B) $= \pi^2/(8(\pi - 1))$ which is 57.6 % in case of a single FET cell and up to $\eta_D$ (Class-B) $= \pi/4$ which is 78.5 % for a push-pull configuration. Nevertheless, Class-B amplifiers are less linear than their Class-A counterparts [82].

**Class-AB**

Class-AB amplifiers are a compromise between Class-A and Class-B operation. They are biased in between Class-A and Class-B operation so that the drain current flows less than 360 degrees, but more than 180 degrees, of each RF cycle. The DC current and drain efficiency are valued between those of Class-A and Class-B setting as shown in Figure 3.4 and Table 3.1.

**Class-C**

An additional efficiency improvement can be achieved by further reduction of the quiescent bias current, as it is the case for Class-C operation. The DC current in Class-C condition is below the threshold current as depict in Figure 3.4. Therefore, the transistor is switched on for only less than half a duty cycle. Even though the resulting drain efficiency is very high (theoretically up to 100 %), Class-C amplifiers have the disadvantage of low output power and power gain leading to a poor PAE due to the low current conduction angle $\alpha$.

There are three more main operating classes (Class-D, -E and -F) for switched amplifiers with which efficiencies up to 100 % are theoretically possible. An excellent review of these conditions can be found in [33]. Examples of Class-F PA design aspects and approaches are reported in, e.g. [37], [89], [92].

Nevertheless, the optimum choice of the operating class depends on the application area and the available technology. The high-power amplifiers discussed in this work are designed for pulsed (FMCW)-Radar applications with the requirement for high output power and maximum power-added efficiency at the same time. On the one hand, Class-A operation offers a poor efficiency and therefore a bad thermal characteristic. On the other hand, Class-C operation delivers only low output power and small gain values. Therefore and for least RF and DC design effort, a Class-AB to Class-B operation close to the maximum transconductance $g_{m,peak}$ for maximum amplification is chosen for all designs within this work. This results in a good tradeoff between output power and efficiency as well as largest RF gain and widest possible bandwidth [109].

### 3.1.3 Large-Signal Modeling

The large-signal (LS) device modeling is not part of this thesis but relevant for an amplifier's circuit design. In particular, the large-signal HEMT models are especially used in the design steps concerning the device selection and the large-signal simulations. The large-signal GaN HEMT models that are used in this work are based on an in-house two-dimensional voltage-lag model to accurately describe thermal effects and low-frequency dispersion, and their impact on the large-signal performance, i.e. the output power, power gain, and PAE. Following the general theory presented in [106], long term memory effects are described by internal states. The parameters of this model are extracted from (pulsed)-DC, small-signal scattering (S)-parameters. Model verification is done using time-domain measurements including the harmonics. This model allows a state-dependent description of the intrinsic drain current and the gate-source and gate-drain space charges. The state quantities are the low-pass filtered voltages $V_{GS}$ and $V_{DS}$. This approach facilitates a correct description of retarded responses (such as the DC characteristics) as well as the instantaneous responses relevant for RF power performance. For the MMIC design, a library of passive microstrip components is available, including all technology specific elements such as MIM capacitors, (spiral) inductors and resistors.

### 3.1.4 Determination of Optimum RF Operating Conditions

The optimum load impedance of a HEMT device depends on the frequency, the operation class determined by $V_{DS}$ and $V_{GS}$, and the input power level. Hereby, the load impedance has a strong impact on the device's characteristics, i.e. the obtainable output power, power gain, and PAE. Based on the requirements of the amplifier, the load resistance $R_L$ can be chosen smaller or larger than $R_{L,opt}$. For example, maximum output power can be obtained simultaneously by maximizing voltage and current swings. According to Figure 3.4, with a Class-A amplifier the optimum load resistance $R_{L,opt}$ results in

$$R_{L,opt}\,(\text{Class-A}) = \frac{\Delta V}{\Delta I} = \frac{V_{max} - V_k}{I_{max}} . \tag{3.13}$$

The most popular techniques to determine the optimum load impedance are the Cripps method as described in [24], load-pull (LP) measurements and nonlinear large-signal simulation of the transistor model. All of these techniques are reviewed in detail in [130].

In this work, the optimum load impedance of a GaN HEMT device is determined via load-pull simulations of the LS-HEMT model. The configuration of the LP simulation of the active device and its simplified output circuit are displayed in Figure 3.6. Here, the FET

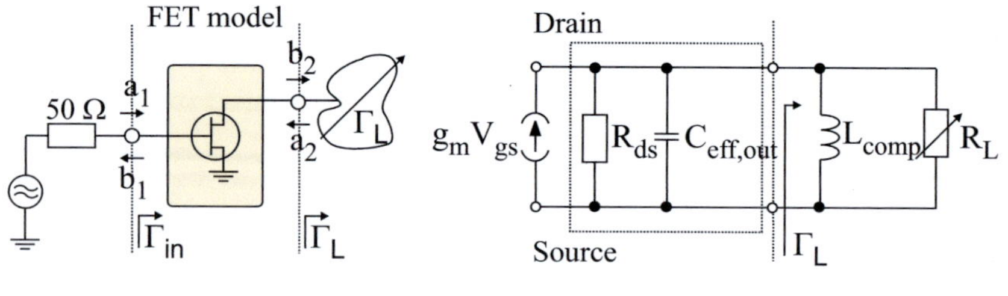


Figure 3.6: Configuration of an AlGaN/GaN-HEMT Load-Pull simulation and simplified active device output circuit.

is connected to a 50 Ω RF source at the input and with a variable output-load circuit.

The equivalent output circuit (figure on the right) of the GaN HEMT can be approximated with the parallel-circuit of the drain-source resistance $R_{ds}$ and $C_{eff,out}$ with

$$C_{eff,out} \approx C_{ds} + C_{gd} \approx 0.4\,\mathrm{pF/mm} \tag{3.14}$$

as a typical value for a 0.25 $\mu$m AlGaN/GaN HEMT technology as considered here. Since $R_{ds} \gg R_L$ in the $I_{DS}$ saturation region it can be neglected. Thus, the load impedance $\Gamma_L$ depends only on $C_{eff,out}$ and $G_L = 1/R_L$ leading to the following relationship

$$\Gamma_L = \frac{a_2}{b_2} = \frac{Y_L - Y_{Load}}{Y_L + Y_{Load}} \quad \text{with} \quad Y_{Load} = G - j\omega C_{eff,out} \tag{3.15}$$

Accordingly, the input impedance $\Gamma_{in}$ that is mainly dependent on the gate-source capacitance $C_{gs}$ and the losses $R_{eff} \approx R_g + R_{gs} + R_s\beta$ with the current amplification factor $\beta = |I_{ds}/I_{gs}| = \left|(g_m V_{gs})/(j\omega C_{gs,eff} V_{gs})\right|$ can be calculated as follows:

$$\Gamma_{in} = \frac{b_1}{a_1} \;\widehat{=}\; R_{eff} + 1/(j\omega C_{gs,eff}) \,. \tag{3.16}$$

Within the LP-simulations the complex output load $\Gamma_L$ of the device is varied to sweep the entire Smith chart or a part of the $\Gamma_L$ plane on a rectangular (real+imaginary) grid for the evaluation of the optimum load location for maximum output power ($\Gamma_{L,Pout}$) and maximum PAE ($\Gamma_{L,PAE}$) as shown in form of green lines in Figure 3.7. In the general case, all harmonic loads above the fundamental are set to 50 $\Omega$. Studies on the influence of a second harmonic termination are given in Section 6.2. In this work, load-pull simulations of the large-signal-HEMT model are performed for Class-AB condition at about 3 dB gain compression over the entire X-band frequency range. Hereby, it is important for the HPA design that the load impedance predicted by the transistor model and the measured value are in good agreement [120]. Since the calculated load impedances $\Gamma_{L,Pout}$ and $\Gamma_{L,PAE}$ of a Class-AB AlGaN/GaN HEMT at X-band frequencies differ strongly from each other, the advantages of the obtainable maximum output power versus the achievable optimum PAE have to be weighted up before starting the actual HPA design.

Consequently, the first step of the design of a high-power-amplifier design is the determination of the optimum RF operating condition of the HEMT device, at first at a single frequency $f_0$. The optimum RF operating condition is specified from the choice of the bias condition

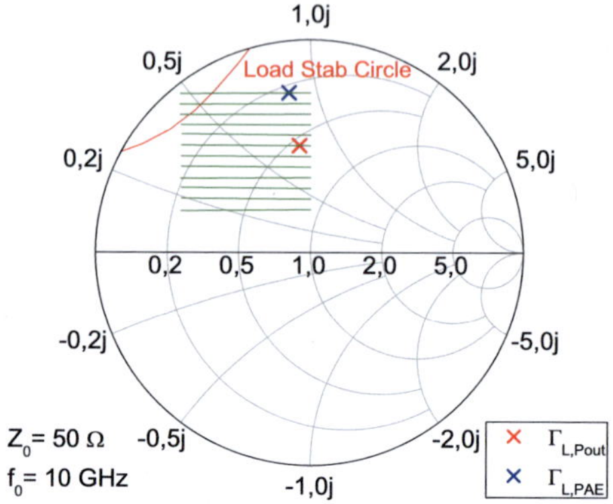


Figure 3.7: Simulated optimum load impedance $\Gamma_{\mathrm{L}}$ for maximum output power and PAE of a 1 mm AlGaN/GaN HEMT at 10 GHz and $P_{\mathrm{in}}$=32 dBm and at $V_{\mathrm{DS}}$=30 V and $V_{\mathrm{GS}}$=-2.25 V (Class-AB operation).

determined by $V_{\mathrm{DS}}$ and $V_{\mathrm{GS}}$, the frequency $f_0$, and the input power level.

Based on the optimum RF operation condition, the optimum load location ($\Gamma_{\mathrm{L,opt}}$) and accordingly the associated optimum input impedance $\Gamma_{\mathrm{in,opt}}$ are defined. Though, the selection of or compromise between the optimum load impedances $\Gamma_{\mathrm{L,Pout}}$ and $\Gamma_{\mathrm{L,PAE}}$ needs to be analyzed based on the HPA design requirements.

$\Gamma_{\mathrm{L}}$ and $\Gamma_{\mathrm{S}}$ equal $\Gamma^*_{\mathrm{in}}$ are realized by means of passive output and input matching networks that are designed to transform $\Gamma$ equal zero to $\Gamma_{\mathrm{L,opt}}$ or $\Gamma_{\mathrm{S}}$. In addition, the input matching network transforms $P_{\mathrm{in,netto}}$ at the gate of the FET nearly lossless to $P_{\mathrm{brutto}}$ at the 50 Ω source.

After a single-frequency analysis, the frequency dependent $\Gamma_{\mathrm{L,opt}}$(f) and $\Gamma_{\mathrm{S,opt}}$(f) are investigated. The main challenge of a matching-network design with the impe-dance $\Gamma_{\mathrm{L}}$(f) consists in the matching to $\Gamma_{\mathrm{L,opt}}$(f) because $\Gamma_{\mathrm{L,opt}}$(f) rotates counterclockwise with increasing frequencies and, in contrast, $\Gamma_{\mathrm{L}}$(f) rotates clockwise. Thus, the frequency-dependent transfer functions are not causal, i.e. they break the Kramers-Kronig relation. As a result, a optimum matching network with minimum losses can only be designed for a single frequency.

The high ohmic load impedance level of the LP simulations of an

AlGaN/GaN HEMT is verified by LP measurements of the same device. It can be estimated according to the above stated simplification with $C_{\text{eff,out}} \approx 0.4\,\text{pF/mm}$ (Equation 3.14) for a non-field-plated AlGaN/GaN 0.25 $\mu$m HEMT technology. This approximation is valid up to frequencies of at least $f_T/3$. Based on the 1 mm total gate width of the device, the obtained $C_{\text{eff,out}}$ value of about 0.4 pF/mm is comparable to $C_{\text{eff,out}}$ values in GaAs 0.25 $\mu$m pHEMT technology where, e.g., $C_{\text{eff,out}} = 0.36\,\text{pF/mm}$ could be obtained in [121]. Otherwise, the ohmic part $R_L$ of the GaN device is about 3-5 times higher than that of the GaAs HEMT together with comparable drain currents and 3-5 times higher drain voltages resulting in a higher obtainable output power which verifies the well-known advantage of the high impedance level of GaN.

## 3.2 Matching Network Design

After the selection of an appropriate operating class and the optimum load impedance of the AlGaN/GaN HEMT depending on the HPA specifications, the passive matching networks at the input and output of each active device stage including the DC supply paths can be designed. The design of an amplifier implies a suitable synthesis of all passive matching networks, accomplishing design requirements and stability issues [41].

For the system description of the circuit design, scattering parameters are used to explain the common power waves as it is usual in high-frequency technologies. The definition of S-parameters is clearly stated in microwave textbooks, e.g. in [69], and therefore will not be described in more detail in this work.

Based on the device models and circuit schematics, CAD (Computer Aided Design) simulations and optimizations are performed using the software tool ADS (Advanced Design System) by Agilent Technologies. For the design of the matching networks, all passive structures such as lines, junctions, MIM capacitors, inductors, and NiCr resistors are available in a microstrip ADS design kit for SiC substrates.

The block diagram shown in Figure 3.8 demonstrates the simplest circuit configuration of a single-stage amplifier, where the circuit consists of a single active device and the input and output fundamental matching networks (IMN and OMN), which include the DC biasing, but no specific harmonic termination networks. For a dual-stage amplifier an additional interstage matching network (ISMN) between the two transistor stages is required.

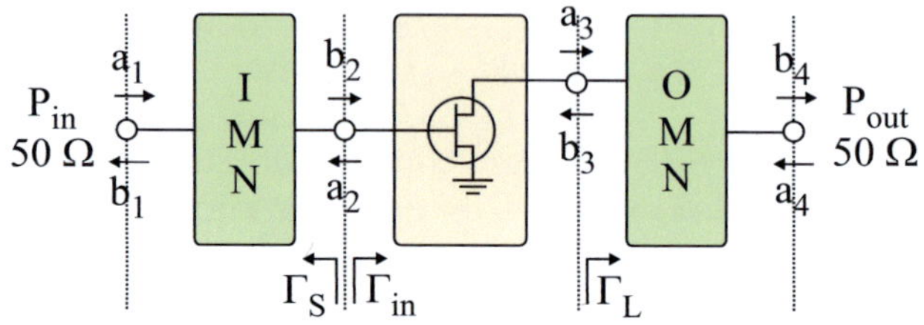


Figure 3.8: Block diagram of the design of a single-stage power amplifier.

Independent of the type of matching network, the basic functions of all passive matching networks are similar. They all consist of

- a splitting/combining network in case of several parallel operated FET cells combined in the transistor-stages
- gate/drain bias voltage supply path(s) to the gates/drains of the transistors
- an impedance transformation network with minimum net insertion loss and
- stabilization elements/circuits for HPA stabilization issues.

According to the HPA specifications, the main task of designing a high-power amplifier is to produce the required output power that mainly depends on the output power density of the HEMT device, the number of transistors that are used in parallel in the transistor stage (or rather PA-stage in case of a dual-stage amplifier), and the net insertion loss of the output matching network. Therefore, the output matching network is the most critical design step relating to the required power performance. Consequently, the design of the HPA is always done from the output to the input of the amplifier with the input matching network being responsible for a uniform gain performance over the required bandwidth.

In general, a specific number of transistors ($N_{FET/DRV}$ and accordingly $N_{FET/PA}$) is used in parallel in the DRV and PA stage of a dual-stage HPA. This makes a low-loss splitting and combining network via microstrip transmission lines at the input and output of each transistor stage necessary. These networks divide the input power equally over all FET cells of the transistor-stage and combine the output power of all power cells of the transistor-stage again. The use of $N_{FET}$ power cells within a transistor-stage necessarily leads to a n-port passive matching network. To simplify the network design and to avoid long computing times, in this work mirror-elements were used within the matching networks to reduce the complexity of the matching networks down to two-port networks with the result of less simulation time.

Another important part of a matching network consists in the gate/ drain bias voltage supply path(s) to the gates/drains of the transistors. Besides delivering bias to the active devices, the bias paths are also used to prevent potential instabilities resulting from a coupling of the various active devices through the bias path. Such coupling can lead to both RF and low-frequency oscillations. Hereby, RF oscillation certainly disrupts the normal HPA operation, but seldom causes a failure of any active device, while a low-frequency oscillation producing large bias-voltage swings might induce permanent damage of the active device [105]. In this work, the bias circuits are realized with microstrip lines and large bias decoupling MIM capacitors at the end of each bias path to provide DC blocking. Furthermore, the gate bias paths often include inductors to reduce the required MMIC chip area and additional stabilization series resistors in front of the MIM blocking capacitors for low-frequency damping and gain reduction.

## 3.2.1 Impedance Transformation

Relating to the basic functions of all matching networks, each network contains an impedance transformation, but with different specific main tasks [66]. Hereby, the output matching network is the most critical network concerning the obtainable power performance, while the interstage and input matching networks are responsible for a uniform gain performance over the required bandwidth that can be obtained when designing the IMN and ISMN for best matching at high frequencies.

For a "perfect match" according to Figure 3.8, the output matching network needs to transform the 50 Ω environment ($\Gamma_L = 0$) to the optimum load impedance $\Gamma_{L,opt}(f)$ of the PA-stage transistors over the required frequency range with minimum net insertion losses to realize maximum output power and power-added efficiency. However, a perfect match can only be achieved for a single frequency, but it is impossible over a finite frequency bandwidth due to the non-causal frequency-dependent transfer functions of $\Gamma_L$ and $\Gamma_{L,opt}(f)$ as already stated above. In addition, there are some theoretical limitations on the maximum available matching bandwidth that were first analyzed by Bode [11] and later expanded by Fano [34]. A detailed description on the bandwidth limitations can be found in [105]. Based on the high $\Gamma_{L,opt}$ of an AlGaN/GaN HEMT at X-band frequencies, a low-loss output impedance transformation is realizable, but only with a poor output matching $S_{22}$ capability.

Furthermore, the input matching networks within this work are designed to match the DRV-stage transistors under conjugate matching conditions ($\Gamma_S(f) = \Gamma_{in}^*(f)$) to the required 50 Ω external load at the input of the amplifier over a given bandwidth. In this work, the optimum $\Gamma_{in}(f)$ is defined based on the before specified $\Gamma_{L,opt}(f)$. Otherwise, a $\Gamma_{in}(f)$ obtained by LP simulations of the FET with output matching network would be slightly improved. However, due to the causality problem concerning the frequency-dependent $\Gamma_{in}(f)$ characteristic, the more precise $\Gamma_{in}(f)$ can not also be exactly realizable.

In addition, the impedance matching on the input side is more challenging than in the OMN case because of the much lower optimum input impedance of the transistors compared with the load impedance resulting in higher net losses. Hereby, a perfect matching of the IMN always comes at the expense of obtainable gain.

In contrast to the OMN and IMN with a matching into a 50 Ω environment on the one side, the ISMN is more complex. It has to transform the optimum load impedance $\Gamma_{L,opt}(f)$ of the DRV-stage transistors into the conjugate optimum input impedance of the PA-stage power cells whereby both impedances are frequency dependent and include reactive parts. Just as for the IMN, the interstage impedance transformations of this work were designed for best matching at the upper end of the given bandwidth with as less net insertion losses as possible for a sufficient flat gain response over the required bandwidth.

The obtainable bandwidth and the achievable power, gain and PAE performance of the entire HPA strongly depend on the number of transformation elements used in the matching networks. With increasing number of elements in the passive circuits, the possible bandwidth of the HPA can be increased as well, but at the expense of higher net insertion losses which in turn lead to an inferior HPA power and PAE performance.

### 3.2.2 Net Insertion Loss

The net insertion loss of each matching network can be calculated individually depending only on the number of used transistors, the S-parameters and the loading of the transistors. These factors and their relationships are defined for the IMN and OMN in Figure 3.8.

Assuming $a_4 = 0$, the net insertion loss of the OMN $L_{OMN}$ in the individual large-signal operating condition of the amplifier depends on the number of transistors used in the PA-stage $N_{FET/PA}$ and can be calculated with $b_4/a_3 = S_{43}$ and $b_3/a_3 = S_{33} = \Gamma_{L,opt}$ as follows:

$$\begin{aligned} L_{OMN} &= \frac{1}{N_{FET/PA}} \cdot \frac{P_{out,OMN}}{P_{in,OMN}} \\ &= \frac{1}{N_{FET/PA}} \cdot \frac{|b_4|^2}{|a_3|^2 - |b_3|^2} \\ &= \frac{1}{N_{FET/PA}} \cdot \frac{|S_{43}|^2}{1 - |S_{33}|^2} \ . \end{aligned} \tag{3.17}$$

The resulting IMN loss $L_{IMN}$ is defined as

$$\begin{aligned} L_{IMN} &= N_{FET/DRV} \cdot \frac{P_{out,IMN}}{P_{in,IMN}} \\ &= N_{FET/DRV} \cdot \frac{|b_2|^2 - |a_2|^2}{|a_1|^2 - |b_1|^2} \ . \end{aligned} \tag{3.18}$$

Taking into account that

$$\begin{aligned} b_2 &= S_{21}a_1 + S_{22}a_2 \\ &= S_{21}a_1 + S_{22}\Gamma_{in}b_2 \\ &= \frac{S_{21}}{1 - S_{22}\Gamma_{in}} \cdot a_1 \end{aligned}$$

$$\begin{aligned} b_1 &= S_{11}a_1 + S_{12}a_2 \\ &= S_{11}a_1 + S_{12}\Gamma_{in}b_2 \\ &= (S_{11} + S_{12}\Gamma_{in}\frac{S_{21}}{1 - S_{22}\Gamma_{in}}) \cdot a_1 \end{aligned}$$

and $b_1 / a_1 = \Gamma_{11}$, $L_{IMN}$ can be restated to

$$\begin{aligned} L_{IMN} &= N_{FET/DRV} \cdot \frac{\frac{|S_{21}|^2}{|1-S_{22}\Gamma_{in}|^2} \cdot (1 - |\Gamma_{in}|^2)}{1 - |\Gamma_{11}|^2} \\ &= N_{FET/DRV} \cdot \frac{\frac{|S_{21}|^2}{|1-S_{22}\Gamma_{in}|^2} \cdot (1 - |\Gamma_{in}|^2)}{1 - \left|S_{11} + S_{12}\Gamma_{in}\frac{S_{21}}{1-S_{22}\Gamma_{in}}\right|^2} \end{aligned} \quad (3.19)$$

dependent only on the number of transistors used in the DRV-stage $N_{FET/DRV}$ and the IMN S-parameters and the loading $\Gamma_{in}$ on port 2. According to the calculated net insertion loss of the input matching network (Equation 3.19), the performance of the IMN can be optimized by minimizing the reflection $\Gamma_{11}$ at the input of the amplifier. For $\Gamma_{11}$ tending towards zero, the net insertion loss becomes lowered.

Finally, the loss of the 2-port interstage matching network $L_{ISMN}$ results in

$$\begin{aligned} L_{ISMN} &= \frac{N_{FET/PA}}{N_{FET/DRV}} \cdot \frac{P_{out,ISMN}}{P_{in,ISMN}} \\ &= \frac{N_{FET/PA}}{N_{FET/DRV}} \cdot \frac{|b_2|^2 - |a_2|^2}{|a_1|^2 - |b_1|^2} \\ &= \frac{N_{FET/PA}}{N_{FET/DRV}} \cdot \frac{\frac{|S_{21}|^2}{|1-S_{22}\Gamma_{in}|^2} \cdot (1 - |\Gamma_{in}|^2)}{1 - |\Gamma_{11}|^2} \quad . \end{aligned} \quad (3.20)$$

As it is commonly used, the net insertion loss $L_{net}$ is always given in dB within this work. The conversion into dB is carried out using the following equation.

$$L_{net}\,[\mathrm{dB}] = -20\log(\sqrt{L_{net}}) \tag{3.21}$$

Typical $L_{net}$ values for AlGaN/GaN HEMT HPAs within X-band frequencies are about 0.5 dB for the output matching network and $\leq$ 2 dB for each of the other matching networks, the IMN and ISMN, depending on the bandwidth of the amplifier.

### 3.2.3 Stabilization Elements of the Matching Networks

To provide overall stability of a HPA design, some important design aspects within the matching networks are considered within this work. The stabilization elements of the matching networks that are used as standard in this work are listed in the following.

1. Series capacitors $C_s$ are required at the input of the IMN and at the output of the OMN for DC blocking, and are therefore included close to the RF ports.

2. The use of resistors within the matching networks always causes a decrease in the possible gain performance of the entire HPA. Therefore, resistors are avoided in the output matching network to prevent any further power losses. However, within the IMN, resistors are well suited for stabilization issues and impedance matching.

3. Series resistors are mostly used in the gate bias paths for low frequency damping and gain reduction as mentioned above.

4. An additional parallel RC network can be included in the IMN for stabilization reasons as described in detail in Section 3.3.1.

5. To prevent odd-mode oscillations (internal stability) the gate and drain connections of the parallel operated FET cells are connected together through resistors and transmission lines (see Section 3.3.1).

## 3.3 Small-Signal and Stability Analysis

To ensure the functionality of the entire power amplifier including all passive matching networks and the active transistor stages, small-signal simulations and optimizations of the amplifier need to be done after designing the matching networks.

The overall system is characterized uniquely for a drain bias voltage sweep from 0 up to 40 V and a simultaneous gate bias voltage sweep around the pinch-off voltage using the S-matrix, which gives a complete overview of the amplifier's small-signal gain ($S_{21}$), its input and output reflection ($S_{11}$ and $S_{22}$), and the reverse isolation ($S_{12}$) behavior. In this case, the S-parameter matrix will be calculated for the above mentioned drain ($V_{DS}$) and gate ($V_{GS}$) bias voltage sweeps over the desired frequency band to cover all relevant biasing conditions. Hereby, the small-signal stability of the amplifier is simply calculable by the S-parameters dependent on the bias operation and the frequency.

### 3.3.1 Stability Considerations

Stability is a very important criterion for the circuit design, as an unstable circuit can oscillate, whereby its performance can be significantly degraded and appropriate control is no longer possible. Therefore, stability analysis has to be performed to investigate if and under which condition the circuit is stable or not. Unstable circuits need to be stabilized. In other words, stability has to be guaranteed for the entire amplifier (two-port) within the impedance termination range of interest and at least for the required frequency bandwidth, but preferably for all frequencies up to the maximum frequency of oscillation $f_{max}$ [33].

#### External Stability

Using the S-parameters, a helpful and frequently employed criterion to prove unconditional stability of a linear two-port can be derived in form of the Rollett factor $k$, introduced in [93]. The following equation

$$k = \frac{1 - |S_{11}|^2 - |S_{22}|^2 + |\Delta|^2}{2\,|S_{12}\,S_{21}|} > 1 \tag{3.22}$$

with $|\Delta| = |S_{11}S_{22} - S_{12}S_{21}| < 1$ defines unconditional stability with $k = 1$ indicating the boundary between unconditional and conditional stability. Hereby, a k-factor analysis is applicable only on a single-stage amplifier. In a multistage environment, the k-factor analysis is still necessary, but by no means sufficient to guarantee overall stability.

Beside the Rollett factor, an alternative criterion to prove unconditional stability, the $\mu$-factor, is established by M. L. Edwards [31] and defined as

$$\mu = \frac{1 - |S_{11}|^2}{|S_{22} - S_{11}^*\Delta| + |S_{12}S_{21}|} > 1 . \qquad (3.23)$$

$\mu$ indicates the minimum distance between the origin of the Unit Smith Chart and the unstable region. Consequently, for $\mu > 1$ the unstable region is outside of the Smith Chart resulting in unconditional stability. In contrast, the Rollett factor itself cannot give secure prediction about unconditional stability. An additional auxiliary condition such as $|\Delta| < 1$ is necessary and sufficient for unconditional stability of a two-port.

**Maximum Available Gain**

An improvement in the $k$- or $\mu$-factor always comes at the price of reduced maximum available gain (MAG) as can be seen in Equation 3.24. The maximum available gain of a two-port network can be calculated as follows

$$MAG = G_{Tmax}|_{k\geq 1} = \left|\frac{S_{21}}{S_{12}}\right| \left(k - \sqrt{k^2 - 1}\right) . \qquad (3.24)$$

**Maximum Stable Gain**

The maximum stable gain (MSG) is reached when the two-port approaches the boundary between unconditional and conditional stability ($k = 1$). In this case, the maximum available gain tends towards its maximum stable value,

$$MSG = G_{\mathrm{T\,max}}|_{k=1} = \left|\frac{S_{21}}{S_{12}}\right| . \qquad (3.25)$$

Figure 3.9 shows the maximal available and stable gain as well as the stability factor $k$ of an 8x125 $\mu$m AlGaN/GaN HEMT at $V_{GS} = -1.75\,V$ and $V_{DS} = 30\,V$. The kink at a frequency of 12 GHz displays the change between MSG and MAG. Below this frequency the transistor is not unconditionally stable ($k < 1$).

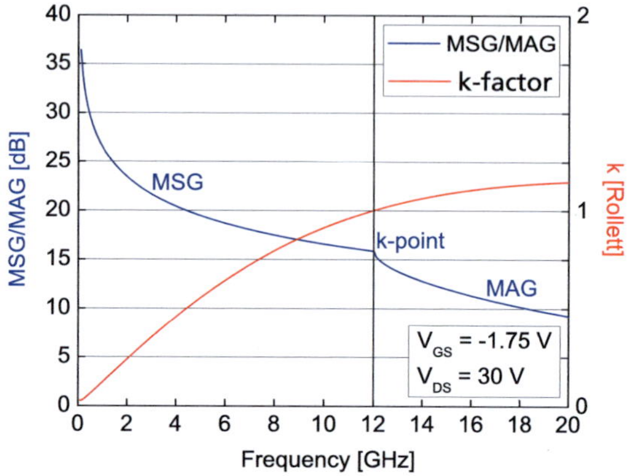


Figure 3.9: MSG/MAG-curve and $k$-factor of a 8x125 $\mu$m AlGaN/GaN HEMT at $V_{GS} = -1.75\,V$ and $V_{DS} = 30\,V$.

**Low-Frequency Oscillations**

The transistors in this work that are designed to operate at microwave frequencies exhibit a very high gain at low frequencies as can be seen in the MSG/MAG curve characteristic of a typical AlGaN/GaN HEMT from the figure above. Therefore, the transistors tend to oscillate at low frequencies. To avoid this low-frequency instability, circuit techniques such as a parallel RC network in the input matching network need to be applied rendering the two-port unconditional stability, but causing a performance sacrifice in form of a decrease in the possible gain [116]. The resistive part of the parallel RC network is used to deliver a stability enhancement by preventing the oscillations at low frequencies. In addition, the usage of series resistors in the gate bias paths provides a low-frequency damping and gain reduction. Resistors in the output matching network should be avoided to prevent power losses.

Nevertheless, there is one important difference between single-stage and dual-stage amplifiers. On the one hand, for dual-stage HPAs unconditional external stability can be obtained relatively easy for $f = 0$ up to $f_{max}$. On the other hand, for single-stage HPAs even with the above mentioned RC stabilization elements, unconditional external stability can be obtained only from a lower frequency limit on. Below this frequency, the HPA is unstable [111]. This is due to the fact that the complex output load $\Gamma_L$ at the transistor's output of a single-stage amplifier always leads to areas where $|\Gamma_{in}| > 1$ for the region where the FET is not unconditionally stable, i.e. for frequencies below a k-factor of 1. In this case, instability can occur. In case of a dual-stage HPA, only a small area of the $\Gamma_{in}$ of the driver stage is covered due to the interstage matching network with the result that normally $|\Gamma_{in}| > 1$ does not occur.

### Internal Stability/Odd-Mode Oscillations

Despite a high external stability of the amplifier system (two-port), the internal stability at the transistor stages has to be analyzed separately. Internal instability occurs in so-called odd-mode oscillations. This can be the case whenever the transistors operated in parallel in each stage are not completely similar due to fabrication/process variations, or where the matching networks are not perfectly symmetrical leading to a feeding mismatch of the single transistors. To avoid such instabilities, a stability analysis using Z-parameter ports directly at the interfaces between the transistor stages and the passive networks is necessary. Ohtomo describes in [73], [72] such a general method of analyzing and simulating stability of multidevice amplifiers with the help of circulators at the interface ports which enable the calculation of the Nyquist plots and voltage distributions of possible instabilities.

To achieve internal stability at the interfaces between the active devices and passive networks, resistors are inserted between the gates and drains of the parallel-operated transistors in each amplifier stage to suppress odd mode oscillations.

If external and internal stability are given for all relevant bias settings over the entire frequency range, a stable amplifier can be obtained according to the small-signal stability analysis. Otherwise, redesigns and optimizations of the matching networks need to be done. Even though a power amplifier can be unconditionally stable under small-signal conditions, it can be unstable under large-signal conditions or vice versa. Thereby, large-signal stability of an amplifier can be a function of the input signal level. Nevertheless, in this work, no additional large-signal stability analysis was applied.

### 3.3.2 Structural Simulation

After the small-signal simulations and stability analysis with ADS, the structures of the circuit should be examined with a 3D structural simulation to identify any coupling effects between the single elements of the passive matching networks. In this work, the simulation software program Ansoft HFSS (High Frequency Structural Simulator) is chosen which is a finite element method solver for electromagnetic structures. Inductors, coupling effects and passive interconnects can be evaluated with this tool. HFSS simulations are done for all matching networks of the amplifier. The results are saved as S-matrices and compared with the ADS small-signal S-parameter simulation results of the entire amplifier.

Mostly a frequency shift down to lower frequencies can be obtained with the structural simulations compared to that obtained with ADS. This frequency shift is frequency dependent, i.e. it is larger for higher frequencies. Thus, this frequency shift seems to support the assumption that it is mainly caused by the "real" capacity value which is not exactly simulated with the MIM capacitor out of the ADS design kit.

## 3.4 Large-Signal Simulation Aspects and Approach

Large-signal analysis is the last design step concerning the different design goals such as output power $P_{out}$, PAE, and power gain under large-signal conditions. This nonlinear analysis is also done in the frequency domain as are the small-signal simulations [127]. In addition, large-signal analysis provides an insight into circuit behavior which cannot be predicted only by small-signal examination. While small-signal characteristics of a circuit design are useful to analyze input feeding with very low power, large-signal simulations are required for circuits operating under input stimuli with high power. A large-signal amplifier must be characterized at all possible input signal levels at least up to output power saturation. Nevertheless, small-signal analysis is suited for first amplifier testing because the calculations are very fast due to the use of simple linear S-matrix calculations.

## 3.5 Schematic to Layout Conversion

After designing the power amplifier in ADS circuit schematic based on device models and the passive structures out of the microstrip ADS design library, the layout of this network must be generated. All circuit schematics of the power amplifier presented in this work are designed in ADS. The corresponding layouts are also generated in ADS, but completed in Cadence Virtuoso where the entire circuit must be controlled with the aid of a design rule check program (DRC) which is adapted to the technology used. Layout optimization is necessary to adjust the power amplifier layout to the desired chip size. Resulting coupling effects between various parts of the network layout must be taken into account. If necessary, re-simulations of the power amplifier need to be done.

## 3.6 Summary of Single Design Steps

The design strategy, presented in this chapter, emphasizes the development of a high-power high-efficiency AlGaN/GaN MMIC amplifier at X-band frequencies. Initially, the design starts with a proper transistor size selection as well as the transistor characterization at high-power operation under optimum load conditions. Thus, the HPAs in this work are biased in Class-AB to Class-B operation as a best trade-off between the obtainable output power and power-added efficiency, regarding the background of radar application.

In a first-step analysis, $\Gamma_{L,opt}$ is determined for a single frequency. Subsequently, $\Gamma_{in}$ is specified only based on the location of $\Gamma_{L,opt}$ in the $\Gamma_L$ plane, but independent of LP simulations of the FET with output matching network since there is only an insignificant discrepancy between the two different methods. A further design step includes a frequency-dependent analysis of $\Gamma_{L,opt}(f)$ and accordingly $\Gamma_{in}(f)$ for X-band frequencies from 8-12 GHz depending on the design requirements. However, the causality problem concerning the frequency-dependent $\Gamma_{L,opt}(f)$ and $\Gamma_{in}(f)$ characteristics result in a non-ideal realization of the frequency-dependent matching networks.

Furthermore, the matching techniques for the different passive matching networks were discussed with the main purpose of maximum power transfer, i.e. minimum net insertion losses, over a required frequency bandwidth. Hereby, the input and load impedances of the AlGaN/GaN HEMT device play a major role concerning the required impedance transformations. The high load impedance of the AlGaN/GaN HEMT enables a low-loss impedance transformation at the output of the device, but leads to a poor output matching $S_{22}$ capability. In addition, the high ohmic load impedance level of the GaN HEMT that is about 3-5 times higher than that of a GaAs device together with comparable drain currents and 3-5 times higher drain voltages results in a higher obtainable output power. The input impedance of the AlGaN/GaN HEMT is much lower, resulting in an IMN with much higher losses than in the OMN.

After the design of the matching networks, the small-signal analysis of the entire HPA was presented including the small-signal gain and matching performances, stability considerations, and 3D structural simulations to detect any coupling effects between the passive elements. In addition, large-signal simulation approaches were given that are important to estimate the achievable output power, power-added efficiency (PAE) and the associated power gain and gain compression of the entire HPA. Finally, the circuit design cycle was completed with the layout design aspect of the power amplifier.

# Chapter 4

# Impact of Technology Progress on Circuit Design

While the previous two chapters describe the RF-performance of an AlGaN/GaN HEMT device and the GaN-related power-amplifier circuit design, this chapter focuses on real AlGaN/GaN HPA MMICs with the aim to investigate the impact of technology progress on circuit designs. Therefore, this chapter gives a short overview of the modified technology aspects and a comparison and evaluation between the measurement results of the corresponding HPA MMICs processed with the varying technology.

## 4.1 Gate Module

In a first study, the influence of the device's gate configuration on the HPA performance is analyzed by means of four different types of gate modules, i.e. a standard gate, a standard gate with additional gate-connected field plate, a T-gate, and finally a T-gate combined with a source-connected second field plate, named as a shield (SH).

Standard HEMTs are generally limited to drain bias voltages leading to a limited RF voltage and RF output power. The implementation of field modulating plates (FP) on a dielectric layer at the drain side of the GaN HEMT delivers some significant improvements concerning the large-signal performance of the device [70], [3], [15]. On the one hand, the function of the field-plate structure is to suppress the current collapse phenomena and to relax the electric-field concentration on the drain side of the gate edge [97]. On the other hand, the benefits of employing an FP electrode are a significantly improved gate-drain breakdown voltage with the result of an increased power capacity of the AlGaN/GaN HEMT as well as reduced high-field trapping effects [3], [142], [13].

## 4.1.1 Standard Gate vs. Gate-Connected Field Plate

The above stated principal function of the field plate is discussed on the basis of two HPA MMIC designs processed with technology G1: one design containing HEMTs with standard gates, named HPA Tinto, and a second design using HEMTs with standard gates and additional gate-connected field plates (GFP), named HPA Reserva.

### Device Structures within Technology G1

The simplified schematic layer structure of an AlGaN/GaN HEMT with standard gates is shown in Figure 4.1. Here, the gate with a gate length of $l_g = 0.25\,\mu$m is processed first, followed by a silicon nitride (SiN) passivation layer.

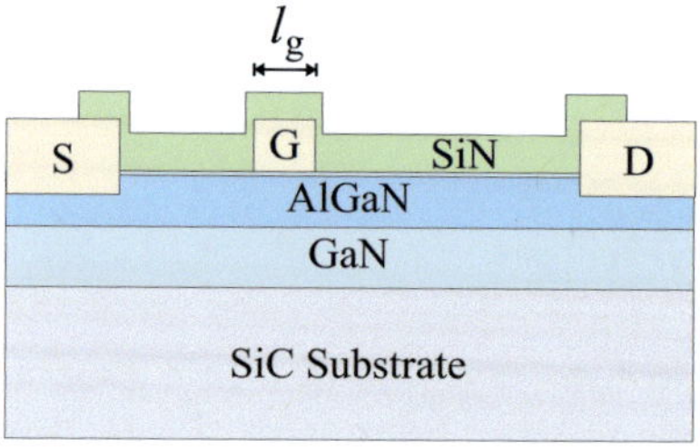


Figure 4.1: Cross-section of an AlGaN/GaN HEMT with standard gates.

Figure 4.2 illustrates the schematic layer structure of an AlGaN/GaN HEMT with standard gates and additional gate-connected field plate on the left side and a focused ion beam (FIB) - scanning electron microscope (SEM) cross-section image of the same device on the right side. As can be seen in both cross-sections, the field plate is deposited

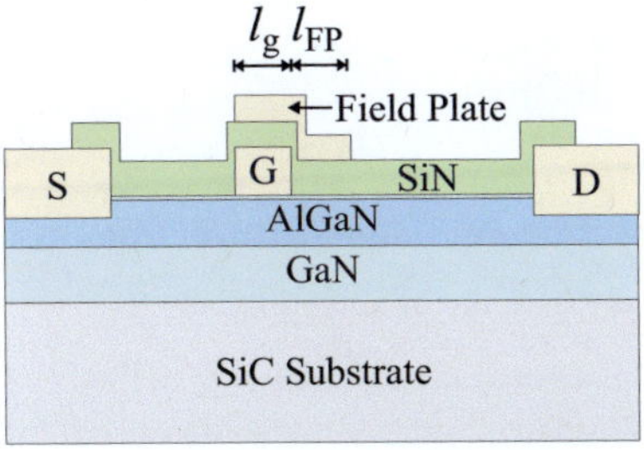


(a) Schematic layer structure

(b) FIB-SEM image ($l_g = 0.5\,\mu$m)

Figure 4.2: Cross-section of an AlGaN/GaN HEMT with standard gates and gate-connected FPs.

on the SiN passivation layer. It is electrically connected to the gate. It's geometry, i.e. the thickness and extension into the gate-drain region $l_{FP}$, as well as the distance above the semiconductor-surface need to be analyzed for optimum performance [74]. In this work, an overlap into the gate-drain region in the order of the gate length $l_g$ is chosen as a best result according to an experimental analysis.

Table 4.1 gives an overview of the impact of the field plate on the intrinsic parameters of the AlGaN/GaN HEMT device. The calculated data that are shown belong to S-parameter measurements of

Table 4.1: Comparison of the calculated intrinsic parameters of 1 mm AlGaN/GaN HEMTs with and without gate-connected field plate measured at $V_{DS} = 30$ V and $I_{DS} = 100 \ldots 150$ mA.

| **Parameter** | **no FP** | **FP** | **Evaluation** relating to HEMT with FP |
|---|---|---|---|
| $C_{gd}$ | 0.07 pF/mm | 0.128 pF/mm | increase of 83 % $\Rightarrow$ decrease in MSG/MAG |
| $C_{gs}$ | 0.93 pF/mm | 1.083 pF/mm | increase of 16 % |
| $C_{ds}$ | 0.202 pF/mm | 0.183 pF/mm | decrease of 10 % |
| $g_m$ | $\approx$ 200 mS/mm | $\approx$ 223 mS/mm | increase of 11 %, but strongly bias-depend |
| $G_{ds}$ | 8.8 mS/mm | 5.5 mS/mm | decrease of 37 % |
| $R_{gd}$ | 23.4 Ωmm | 7 Ωmm | decrease of 70 % $\Rightarrow$ lower feedback losses |

AlGaN / GaN HEMT devices with and without gate-connected field plate operated at $V_{DS} = 30\,V$ and $I_{DS} = 100\,mA$.

### HPA Performance Dependent on the Use of Gate-Connected Field Plates

Both HPA MMICs, Tinto and Reserva, are designed in a similar way for easy comparison. Both designs demonstrate single-stage amplifiers containing a transistor stage and a fundamental input and output matching network (IMN and OMN). Thereby, the transistor stage consists of two parallel operated 1 mm basic AlGaN / GaN-HEMT cells with an $8 \times 125\,\mu m$ gate geometry. Both HPAs are designed for the lower X-band frequency range from 8 up to 9 GHz.

Due to the fact that the additional field plate causes a strong increase in $C_{gd}$ and consequently a higher output capacitance $C_{eff,out}$ (see Equation 3.14), $Y_{L,FP}$ exhibits a relatively higher imaginary part in the admittance plane compared with the optimum output admittance of the device without field plate. The simulated frequency-dependent optimum load reflection coefficient $\Gamma_L$ values of both devices at 9 GHz are $\Gamma_{L,noFP}(9\,GHz) = 0.62\,\angle 104^\circ$ and $\Gamma_{L,FP}(9\,GHz) = 0.76\,\angle 114^\circ$. Consequently, the transformation within the OMN of the HPA Reserva from $\Gamma_L = 0$ to the required optimum load impedance $\Gamma_{L,opt}$ of the transistor output is also larger. Accordingly, the HPA Reserva offers a lower gain performance ($MSG \approx g_m / (\omega C_{gd})$) due to the strong increase in $C_{gd}$.

Figure 4.3 shows the chip images of both single-stage HPA MMICs.

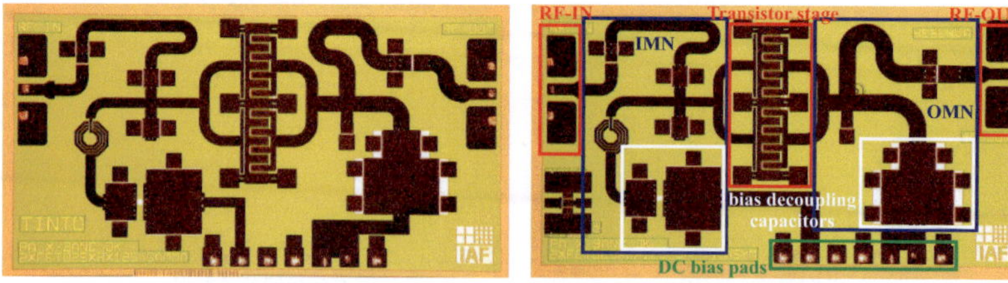


(a) HPA Tinto (b) HPA Reserva

Figure 4.3: Photographs of the HPAs Tinto and Reserva, each with a chip size of $2.75 \times 1.5\,mm^2$.

The HPA Tinto is depicted in Figure 4.3(a) and the HPA Reserva is displayed in Figure 4.3(b), each including large-area bias decoupling capacitors and DC bias pads for direct on-wafer tests. The single parts of the network design are marked exemplarily in the chip image of the HPA Reserva.

**Measurement Results**

All measurements were performed in an on-wafer configuration on the same wafer for both the HPA Tinto and Reserva to avoid any wafer-to-wafer deviation. Thereby, the small-signal S-parameter measurements were done under continuous wave (CW) operation and the large-signal (LS) measurements under pulsed-RF condition at an ambient temperature of $T_a$=25 °C.

Figures 4.4 and 4.5 show the CW-S-parameter results of the HPAs Tinto and Reserva at a biasing of each $V_{\mathrm{DS}}$= 15 V and $V_{\mathrm{GS}}$= -4 V, respectively. Overlays of the measured CW-S-parameters of five HPA MMICs of the type Tinto and Reserva, each from different cells of the same wafer, are depicted in Figure 4.4(a) and 4.4(b), respectively.

The measured S-Parameters of the HPA Tinto are very uniform for the five samples. In contrast, the overlay of the measured S-parameters of the HPA Reserva shows a 2 dB deviation between the different samples that can be explained by the process variation of the additional FP due to the high E-beam lithography alignment requirement of the FP related to the gate.

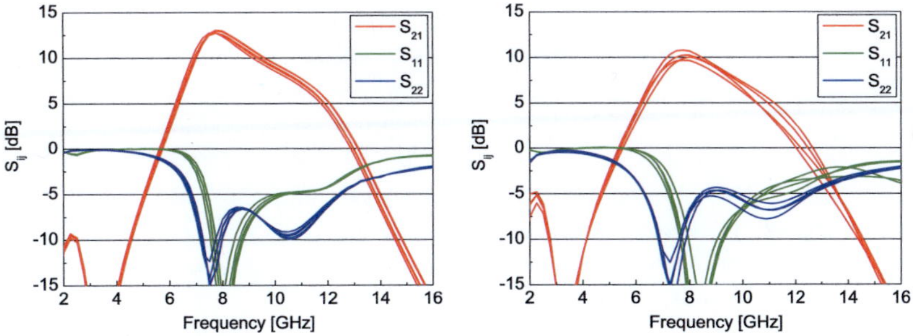


(a) HPA Tinto (b) HPA Reserva

Figure 4.4: Overlay of measured CW-S-parameters of five samples of the type HPA Tinto and Reserva. Bias point: $V_{\mathrm{DS}}$= 15 V and $V_{\mathrm{GS}}$= -4 V.

Figures 4.5(a) and 4.5(b) give a comparison of the simulated and measured CW-S-parameters of the MMIC designs Tinto and Reserva, respectively. While the HPA Tinto shows a measured maximum small-signal gain of 13 dB at 8 GHz, the equivalent maximum gain of the HPA Reserva is 10.2 dB, i.e. 2.8 dB lower than that of its pendant HPA Tinto. The reduced gain performance is the result of the gate-connected FP configuration that offers an additional capacitance between the FP and the drain leading to an increased gate-to-drain capacitance $C_{gd}$, i.e. a high feedback capacitance and hence a much lower MSG than the non-field-plated device [139]. However, the input and output reflections as well as the reverse isolation are comparable for both designs.

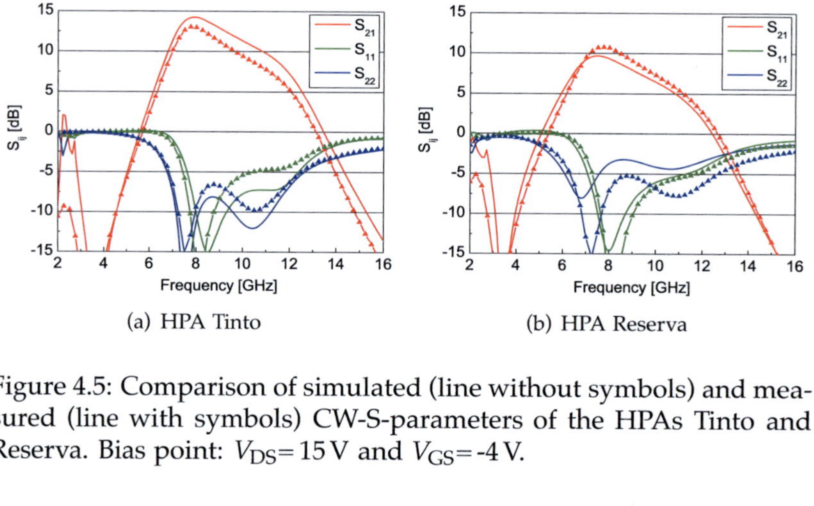


(a) HPA Tinto

(b) HPA Reserva

Figure 4.5: Comparison of simulated (line without symbols) and measured (line with symbols) CW-S-parameters of the HPAs Tinto and Reserva. Bias point: $V_{DS}$= 15 V and $V_{GS}$= -4 V.

The power performances of both MMICs were measured in pulsed-mode with a pulse width and a duty cycle of 100 $\mu$s and 10 %, respectively. The same bias condition of $V_{DS}$= 30 V and $V_{GS}$= -4 V was chosen for both HPAs. Figure 4.6 presents a comparison between the large-signal simulations (line without symbols) and measurements (line with symbols) of the output power, PAE and gain performance of the HPAs Tinto and Reserva as a function of input power levels at 8 GHz. For both HPA MMICs, simulation and measurement are in a very good agreement up to an input power of approximately 26 dBm. For higher input power levels, the simulation is too optimistic.

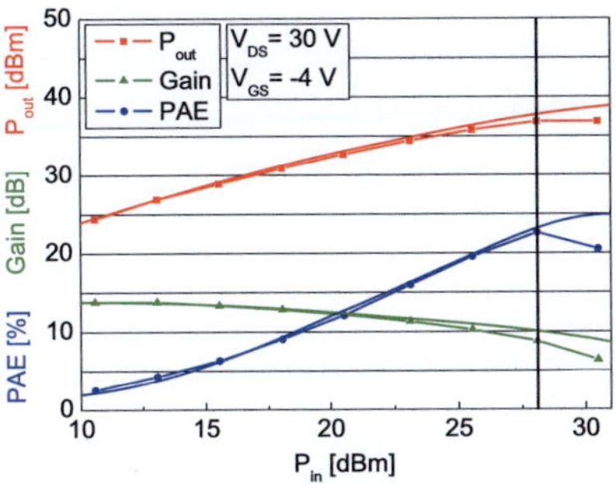


(a) HPA Tinto

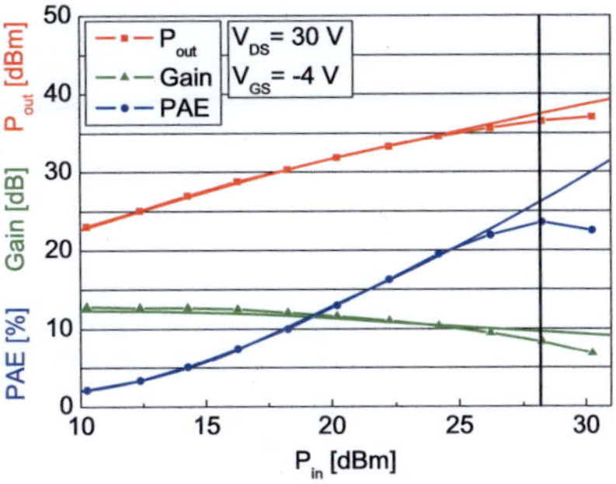


(b) HPA Reserva

Figure 4.6: Large-signal simulations (line without symbols) and pulsed-RF measurements (line with symbols) of the HPAs Tinto and Reserva at 8 GHz. Bias point: $V_{DS}$= 30 V and $V_{GS}$= -4 V.

Nevertheless, the highest PAE values of 22.5 % and 23.5 % were measured on the HPAs Tinto and Reserva respectively at an input power level of 28 dBm as marked in both figures. At the same input power level, the HPA Tinto offers its maximum output power of 36.8 dBm with an associated power gain of 8.8 dB under approximately 5 dB gain compression. For $P_{in} > 28$ dBm, the PAE and gain performances strongly decrease.

The output power and power gain of the power amplifier Reserva yield 36.6 dBm and 8.3 dB at $P_{in} = 28$ dBm and at a reduced gain compression of closely 4 dB, respectively. Nevertheless, in this case the output power is still increasing at $P_{in} = 28$ dBm and reaches its saturated value of 37 dBm only at an input power level of 30 dBm. In addition, the PAE and gain values do not drop as much as for the HPA Tinto for input power levels beyond 28 dBm.

**Discussion**

Table 4.2 gives a comparison of the pulsed-RF-measurement results of the HPA MMICs Tinto and Reserva biased at $V_{DS}$= 30 V and $V_{GS}$= -4 V each. As a result, the HPA Reserva shows a slightly higher output power and a PAE improvement of approximately 1 % compared with the HPA Tinto.

Table 4.2: Comparison of the pulsed-RF-measurement results of the HPAs Tinto and Reserva at $f_0$=8 GHz and at a biasing of $V_{DS}$=30 V and $V_{GS}$=-1.6 V.

| **Main features** at $P_{in}$=28 dBm | **HPA Tinto** "standard gate" | **HPA Reserva** "gate with GFP" | **Evaluation** relating to HPA Reserva |
|---|---|---|---|
| $P_{out}$ | 36.8 dBm ≙ 4.8 W | 36.6 dBm ≙ 4.6 W | about the same |
| $PAE_{max}$ | 22.5 % | 23.5 % | factor 1.05 in PAE |
| power gain | 8.8 dB | 8.3 dB | - 0.5 dB due to higher $C_{gd}$ |
| gain comp. | about 5 dB | < 4 dB | 1 dB improvement |
| $P_{out,sat}$ | 36.8 dBm | 37 dBm | |
| linear gain | 14 dB | 12 dB | 2 dB lower |

Primarily, the HPA Reserva shows a first approach to the expected HPA RF-perfor-mance improvement due to the adoption of gate-connected field plates. Nevertheless, the obtained measurement results are still not optimized as this was the first technology run for the examination of the use of gate-connected field plates. The most important improvements based on the use of GFPs are summarized in the following.

The HPA Tinto shows premature saturation effects compared with the HPA Reserva due to charge trapping effects on the semiconductor surface and RF breakdown in the conducting channel [118]. These effects can be reduced by the use of field plates as can be seen in the pulsed-RF-measurement results of the HPA Reserva offering a slightly increased saturated output power and accordingly a slightly higher PAE. Further PAE and $P_{out}$ improvements are expected with the help of an improved processing technology.

However, a disadvantage of the HPA using HEMTs with gate-connected field plates consists in the reduced linear and power gain performance as a result of the higher feedback capacitance and accordingly lower MSG behavior of the HEMT device using GFPs.

Another important aspect consists in the reproducibility of the HPA designs. The overlay of the measured S-parameters of various HPA MMICs from the type Reserva shows a deviation of about 2 dB between the single samples as can be seen in Figure 4.4(b). This discrepancy results from the process variation of the additional field plates that is strongly affected by the high E-beam lithography alignment requirement of the field plate related to the gate. To overcome this problem, in a further technology-progress step, the gate and the GFPs are processed within one process step.

### 4.1.2 Gate-Connected Field Plate vs. T-Gate

The advantage of using devices with T-gates instead of standard gates with additional gate-connected field plates is explained in this section on the basis of simulation and measurement results of the modified HPA MMIC Reserva, described in detail above, now containing HEMT devices with T-gates. In general, T-gate means that the gate is processed with an integrated field plate that overlaps the gate likewise on the source and drain side [76].

#### Characteristics of T-Gate within Technology G2

Figure 4.2 shows the cross-section of an AlGaN/GaN HEMT with T-gates by means of a schematic layer structure on the left side and a FIB-SEM image on the right side. The principle functions and benefits of the T-gate are comparable to those of the gate-connected field plate.

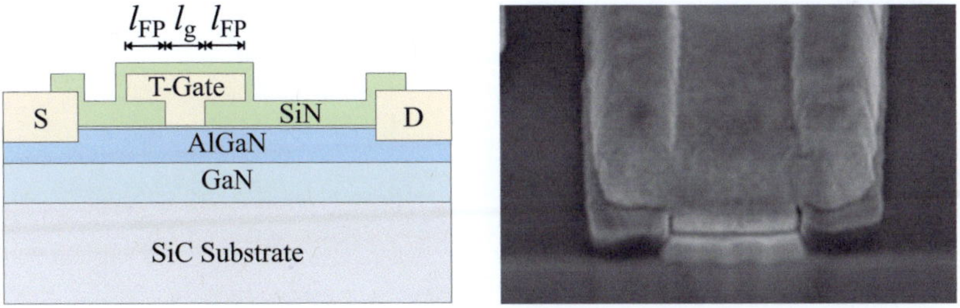


(a) Schematic (b) FIB-SEM image ($l_g = 0.5\,\mu m$)

Figure 4.7: Cross-section of an AlGaN/GaN HEMT with T-gate.

Nevertheless, the main advantage of the T-gate compared with the standard gate with additional gate-connected field plate is the enhanced reproducibility of the T-gate. This is due to the fact that the T-gate is processed within one process step while the gate with an additional field plate needs an additional process step for the field plate leading to further process variation.

Table 4.3 gives an overview of the impact of the use of a T-gate instead of an additional gate-connected field plate on the intrinsic parameters of the AlGaN/GaN HEMT device. The calculated data that are shown belong to S-parameter measurements of AlGaN/GaN HEMT devices with gate-connected field plate and with T-gate operated at $V_{DS} = 30\,V$ and $I_{DS} = 100\,mA$.

Table 4.3: Comparison of the calculated intrinsic parameters of 1 mm AlGaN/GaN HEMTs with gate-connected field plate and T-gate measured at 30 V and $I_{DS} = 100 \ldots 150\,mA$.

| **Parameter** | **FP** | **T-gate** | **Evaluation** relating to the T-gate HEMT |
|---|---|---|---|
| $C_{gd}$ | 0.128 pF/mm | 0.13 pF/mm | nearly unchanged |
| $C_{gs}$ | 1.083 pF/mm | 0.92 pF/mm | decrease of 15 % |
| | | | about the same value as without FP |
| $C_{ds}$ | 0.183 pF/mm | 0.2 pF/mm | increase of 9 % |
| | | | about the same value as without FP |
| $g_m$ | ≈ 223 mS/mm | ≈ 210 mS/mm | in between the values "with/without FP" |
| $G_{ds}$ | 5.5 mS/mm | 9.7 mS/mm | increase of 76 % |
| | | | slightly higher than without FP |
| $R_{gd}$ | 7 Ωmm | 10.3 Ωmm | increase of 47 % |
| | | | in between the values "with/without FP" |

## Advantage of T-Gate over Gate-Connected Field Plate

The improved reproducibility is shown in Figure 4.8. The figure gives a comparison of the overlay of the measured CW-S-parameters of five samples of the HPA Reserva processed with the two different gate topologies. On the left in Figure 4.8(a), the HPA Reserva is processed with transistors containing standard gates with GFPs, named HPA Reserva I. On the right, Figure 4.8(b) shows the measured data of

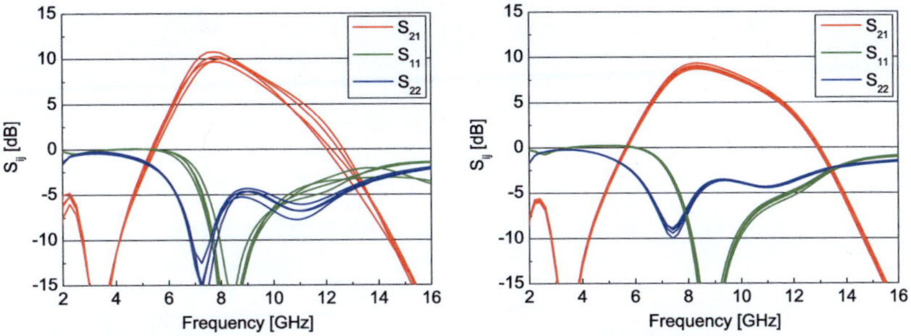


(a) HPA Reserva I (gate with GFP)

(b) HPA Reserva II (T-gate)

Figure 4.8: Overlay of CW-S-parameter-measurements of five samples of the HPA Reserva with different gate structures at $V_{DS}$= 15 V and $I_{DS}$= 150 mA.

the HPA Reserva processed with T-gate HEMT cells, named HPA Reserva II. Even though the maximum small-signal gain of the HPA Reserva I is higher, the overlay of the HPA Reserva II samples offers an improved and very good reproducibility for all small-signal S-parameters compared with the overlay of the HPA Reserva I samples. In this case, the slightly reduced small-signal performance of the T-gate variant is due to the new processing technology and the fact that the integrated field plate of the T-gate structure with slightly changed dimensions compared to the GFP delivers a stronger impact on the HEMT performance resulting in a further decrease of the MSG/MAG behavior.

After the first process run with the T-gate technology, a T-gate HEMT model was developed [106]. Figure 4.9(a) shows the CW re-simulated (line without symbols) and measured (line with symbols) S-parameters of the MMIC Reserva II processed with the T-gate structure. The simulations and measurements were done at $V_{DS}$= 15 V and $I_{DS}$= 150 mA offering a very good agreement for all frequencies. The obtained maximum measured gain was 9.5 dB at a frequency of 8 GHz.

Large-signal simulations were done with the new FET model as well and compared with the pulsed-RF-measurements of the HPA Reserva II at 8 GHz and $V_{DS}$= 30 V. The results are shown in Figure 4.9(b) and demonstrate a good agreement for the $P_{out}$ and gain curves,

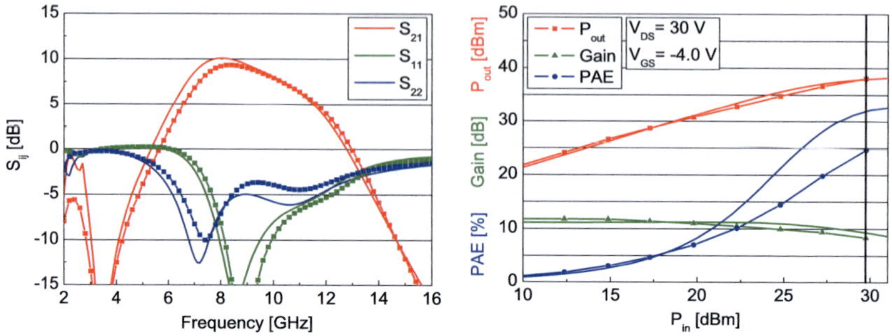


(a) CW-S-parameters at $V_{DS}$= 15 V (b) Pulsed-RF performance at 8 GHz

Figure 4.9: Simulated (line without symbols) and measured (line with symbols) small-signal and large-signal performance of the HPA Reserva II.

but show a too optimistically simulated PAE for input power levels beyond 17 dBm. This discrepancy between simulation and measurement of the PAE curve results from the new T-gate model that was developed only on the basis of the measured S-parameters. Therefore, the simulated and measured small-signal S-parameters fit together very well for all frequencies. However, the large-signal replica is quite bad because the preliminary HEMT-model includes just low-frequency dispersion, but no DC-RF-dispersion modelling.

Nevertheless, a maximum measured PAE and output power were reached for the highest measured input power level of 29 dBm. The trend of the PAE and $P_{\text{out}}$ curves at $P_{\text{in}} = 29$ dBm indicate a tendency for a further increase towards even higher input power levels. Processed with the T-gate HEMT structure, the HPA Reserva delivers an output power and PAE of 38.1 dBm and 25 % at an input power level of 29 dBm, respectively. The associated gain yields 8.5 dB. Compared with the HPA Reserva I, a further increase of the output power and PAE could be obtained with the T-gate technology, i.e. 1.5 dB more output power and accordingly a 1.5 % higher PAE with an unchanged associated power gain.

**Discussion**

The small-signal and large-signal measurement results of the HPA MMICs Reserva I and Reserva II are summarized in Table 4.4.

Table 4.4: Comparison of the pulsed-RF-measurement results of the HPA Reserva processed with standard gates with additional GFPs (HPA Reserva I) and with T-gates (HPA Reserva II) at $V_{DS}$= 30 V and $V_{GS}$= -4 V.

| **Main features** | **HPA Reserva I** "gate with GFP" | **HPA Reserva II** "T-gate" | **Evaluation** relating to HPA Reserva II |
|---|---|---|---|
| **at $f_0$= 8 GHz** | **and $P_{in}$= 28 dBm** | **and $P_{in}$= 29 dBm** | |
| $P_{out}$ | 36.6 dBm $\widehat{=}$ 4.6 W | 38.1 dBm $\widehat{=}$ 6.5 W | + 1.5 dB |
| $PAE_{max}$ | 23.5 % | 25 % | factor 1.07 |
| power gain | 8.3 dB | 8.5 dB | similar |
| gain comp. | 3.7 dB | 3.5 dB | similar |
| $P_{out,sat}$ | 37 dBm | not reached yet | |
| linear gain | 12 dB | 12 dB | no change |
| max. measured small-signal gain at $V_{DS}$= 15 V | 10.5 dB large deviation between diff. samples | 9.5 dB little deviation ⇒ improved reproducibility | - 1 dB because of first process run with T-gates |

Comparing both HPA Reserva variants processed with GFP and T-gate technology, the measured power gain behavior remains nearly stable because there is almost no further change in the feedback capacitance $C_{gd}$.

The implementation of the T-gate instead of a standard gate with additional gate-connected field plate delivers an improved reproducibility as can be seen in Figure 4.8. While there is a large deviation between the different sample measurements of the HPA Reserva I processed with GFP, there is almost no discrepancy in the case of the T-gate variant. The reason for the improved reproducibility of the T-gate technology is that the T-gate is processed within one process step while the gate with additional field plate needs an additional process step for the field plates leading to further process variation.

Moreover, the use of field plates in general and the use of T-gates in particular leads to an increase in the obtainable output power at higher $V_{DS}$ of a HPA MMIC, due to the reduced dispersion effects. Here, for the highest measured input power level of 30 dBm, the output power is still increasing.

Finally, the power-added efficiency can also be further increased by the use of the new T-gate technology. In case of a standard gate without any field plates, a PAE of 22.5 % was reached on design level. This value was increased by 1 % due to the use of additional GFPs. With the new T-gate technology an additional 1.5 % PAE-enhancement was obtained resulting in a maximum PAE of 25 %, i.e. a PAE improvement of a factor of 1.1 in total. These results prove that the use of a field plate delivers the expected positive impact on the PAE behavior of a HPA MMIC design that in turn depends on the field plate geometry.

Due to the advantageous performance of the T-gate HEMT device, this type of transistors is used for all further HPA designs within this work which are discussed in detail in Chapter 5 and Chapter 6.

### 4.1.3 Source-Connected Shield

The last gate topology that is taken into consideration within this study is a T-gate combined with a second field plate, here named a shield (SH). The shield is applied after the passivation of the gate metal and terminated to the source to minimize the feedback capacitance since a high feedback capacitance causes a much lower MSG behavior of the HEMT device. The drawback of additional $C_{gd}$ of the gate-connected field plate hence is more than compensated using a source-connected field plate. A schematic and FIB image of the cross-section of a GaN HEMT with T-gate and source-connected shield is illustrated in Figure 4.10.

The function of the shield is to maintain the maximization of the biasing voltage $V_{ds}$ at large-signal operation without additional feedback capacitance. Thus, source-connected field plates offer a significant improvement in maximum available and stable gain compared with GFP and non-FP devices [70], [141] resulting in higher efficiencies. In addition, a further reduction of electrical field peaks is expected leading to even higher breakdown voltages compared with the GFP HEMT device.

However, the field-plate variation in the GaN technology mainly influences the capacitance matrix of the GaN HEMT. As already mentioned above, $C_{gd}$ is comparatively lower for the device with shield than for a transistor without shield independent of the use of GFPs. In addition to the decrease in $C_{gd}$, $C_{gs}$ increases resulting in a lower broadband matching capability. Besides the change in the capacitance

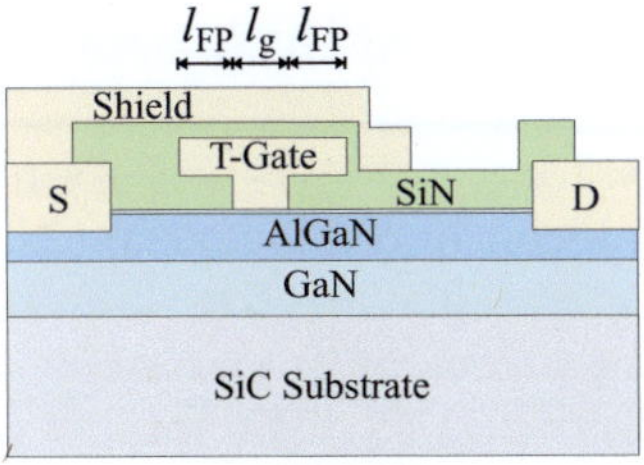


(a) Schematic

(b) FIB-SEM image ($l_g = 0.5\,\mu$m)

Figure 4.10: Cross-section of an AlGaN / GaN HEMT with T-gate and with a source-connected shield.

matrix, the source-connected field plate causes a significant increase in the gate-drain space charge region resistance $R_{gd}$ due to slow-down of the charge carrier which results in higher feedback losses and analogous to a decrease in PAE. A further analysis of the impact of a source-connected shield on a T-gate HEMT device of the Fraunhofer IAF AlGaN / GaN technology is given in [57]. Table 4.5 gives an overview of the impact of the use of a source-connected shield on the intrinsic parameters of the AlGaN / GaN HEMT device. The calculated data that are shown belong to S-parameter measurements of

Table 4.5: Comparison of the calculated intrinsic parameters of 1 mm AlGaN / GaN HEMTs with and without source-connected shields measured at 30 V and $I_{DS} = 100 \ldots 150$ mA.

| **Parameter** | **without SH** | **with SH** | **Evaluation** relating to HEMT with SH |
|---|---|---|---|
| $C_{gd}$ | 0.1 pF / mm | 0.07 pF / mm | decrease of 30 % $\Rightarrow$ increase in MSG / MAG |
| $C_{gs}$ | 0.95 pF / mm | 1.12 pF / mm | increase of 20 % $\Rightarrow$ lower broadband matching capability |
| $C_{ds}$ | 0.21 pF / mm | 0.27 pF / mm | increase of 35 % |
| $g_m$ | $\approx$ 229 mS / mm | $\approx$ 236 mS / mm | nearly constant, but strongly bias-dependent |
| $G_{ds}$ | 5.9 mS / mm | 7 mS / mm | increase of 19 % |
| $R_{gd}$ | 10.4 $\Omega$mm | 19.3 $\Omega$mm | increase of 86 % |

T-gate AlGaN / GaN HEMT devices with and without shield operated at $V_{DS} = 30$ V and $I_{DS} = 100$ mA.

The Fraunhofer IAF AlGaN / GaN HEMT technology using source terminated field plates was analyzed and recently published in [132, 57, 59]. For a 1 mm TGW HEMT device operated at 10 GHz and at a biasing of 30 V, mean values for the PAE, $P_{out}$, and linear gain of 50 %, 38 dBm, and 13 dB, respectively were obtained [132]. The impact of the shield on the transconductance $g_m$ and MSG / MAG performance of this GaN HEMT is described in [59].

### HPA Design Aspects

To analyze the impact of the use of an additional source-connected field-plate device on a HPA circuit design, this study compares two dual-stage HPA designs using FET devices without (HPA Cesar V1) and with (HPA Cesar V2) shields. Both MMICs were designed in the style of the HPA Sacy, see Chapter 6.1. They both offer the same PA / DRV gate-width ratio of 3:1 as the MMIC Sacy, but with a smaller total gate-width adapted to the design goals of a satcom project for telemetry applications in space addressed with these designs. Therefore, both HPAs Cesar V1 and Cesar V2 were measured in CW- and not in pulsed-RF power operation.

The current design goals for the design variants Cesar V1 and Cesar V2 include a high power and high PAE performance at a low compression level of -2 dB in CW-operation at a drain voltage of $V_{DS}$=28 V. Here, an output power of approximately 37 dBm and a PAE above 35 % with an associated gain $\geq$ 20 dB is aimed at a 2 GHz bandwidth from 8 GHz up to 10 GHz. The resulting core characteristics of both HPA MMICs Cesar V1 and Cesar V2 are summarized in Table 4.6.

Table 4.6: Design characteristics of the HPA MMICs Cesar V1 and V2.

| MMIC | PA/DRV GW Ratio | T-Gate Structure | Content of Al | source-connected shield |
|---|---|---|---|---|
| Cesar V1 | 3:1 | 2×8×125 μm (PA-stage)<br>1×8×85 μm (DRV-stage) | 22 % | no |
| Cesar V2 | 3:1 | 2×8×125 μm (PA-stage)<br>1×8×85 μm (DRV-stage) | 22 % | yes |

The matching networks of design Cesar V1 and Cesar V2 slightly differ from each other, because the use of shields within the HEMTs of the HPA Cesar V2 leads to only small changes in the input and output impedance of the transistors. This is because the increase in $C_{ds}$ approximately compensates for the decrease in $C_{gd}$, i.e. $C_{eff,out} \approx C_{ds} + C_{gd}$ remains nearly constant for both device variants. In both HPAs, the matching networks are designed for $\Gamma_{L,PAE}$ due to the challenging PAE design goal. Since both designs are similar, only the chip image of the HPA Cesar V1 is shown in Figure 4.11.

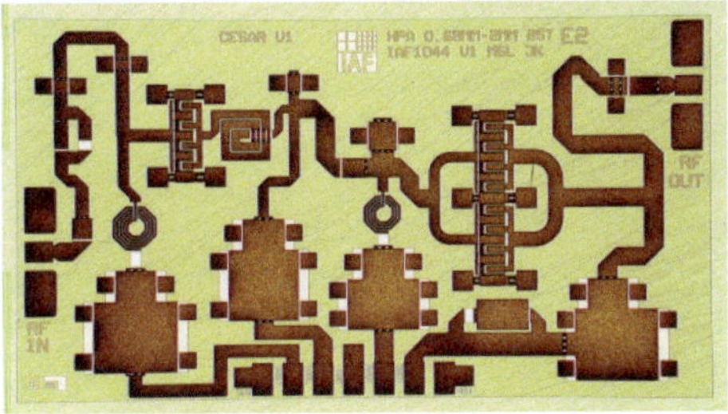


Figure 4.11: Photograph of the dual-stage HPA Cesar V1. The chip size is $3.5 \times 2\,\text{mm}^2$.

### Impact of the Shield on the HPA Performance

Small signal and large signal measurements were performed in an on-wafer configuration at a drain voltage of $V_{DS}$= 28 V and $I_{DS}$= 100 mA in CW-operation. All data of the HPA MMICs Cesar V1 and Cesar V2 shown below were taken from the same wafer to mitigate any variations. Thereby, the CW-S-parameters were found to be very uniform over the entire wafer.

Figure 4.12(a) shows a comparison of the measured S-parameters of the HPAs Cesar V1 (line with symbols) and Cesar V2 (line without symbols). Both HPAs offer a small-signal gain $\geq$ 20 dB for a frequency range from about 6 GHz up to 10 GHz. Within this frequency range, the curve progression of the HPA Cesar V2 is more constant and about 2 - 6 dB higher than that of the HPA Cesar V1. Furthermore, both curves show a sudden decrease of gain at the upper frequency band edge. While the small-signal gain of HPA Cesar V1 starts dropping rapidly at 9.5 GHz, the gain curve of MMIC Cesar V2 starts falling at a 0.7 GHz higher frequency of about 10.2 GHz.

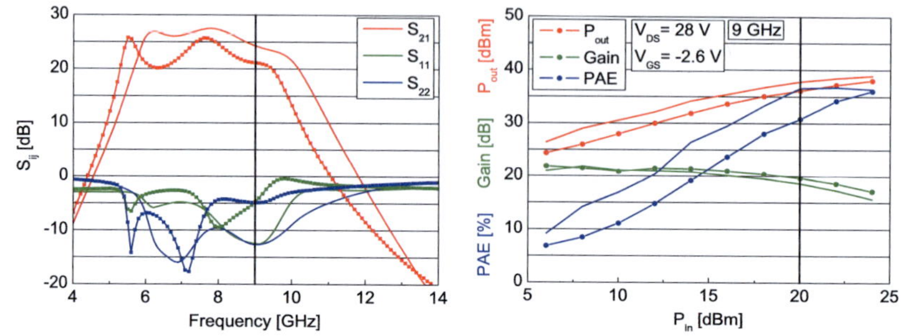


(a) CW-S-parameters (b) CW-power-measurements at 9 GHz

Figure 4.12: Small-signal and large-signal measurements of the HPA MMICs Cesar V1 (line with symbols) and Cesar V2 (line without symbols) each at $V_{DS}$=28 V.

Looking at the input and output reflection $S_{11}$ and $S_{22}$, the shield variant shows a significant improvement of the matching capability around the center frequency of 9 GHz. The HPA Cesar V1 has an input and output reflection of about -5 dB at the center frequency, whereas the HPA Cesar V2 exhibits an input and output matching of approximately -12 dB at 9 GHz and of better than -5 dB for the frequency range from 6 GHz up to 10 GHz.

To sum up, the shield variant delivers a better overall small-signal performance concerning the small-signal gain and matching capabilities over the entire frequency range, but primarily an improvement towards higher frequencies.

Large-signal measurements were also done in CW-operation at the same bias conditions. The power-sweep results of both MMIC designs at the center frequency of 9 GHz are shown in Figure 4.12(b). Again, the HPA Cesar V1 is represented by the use of lines with symbols and the HPA Cesar V2 by the use of lines without symbols.

Both MMICs offer the same linear gain of about 21-22 dB leading to a -2 dB gain compression with an associated power gain of about 20 dB at an input power level of $P_{\text{in}} = 20$ dBm as marked in the Figure. At this input power level, the resulting gain values are 20 dB and 19 dB for the MMICs Cesar V1 and Cesar V2, respectively. The required gain of $\geq$ 20 dB at a -2 dB compression level was just about reached with both the HPA Cesar V1 and the HPA Cesar V2 at 9 GHz.

Compared with the HPA Cesar V1, the MMIC Cesar V2 offers a significantly improved PAE performance up to an input power level of 23 dBm and a nearly constant 2 dB increase of $P_{\text{out}}$ over the entire input power range. As a result, the HPA Cesar V2 yields 38 dBm output power and 37 % PAE at the required -2 dB compression level, while the MMIC Cesar V1 provides only an output power and PAE of 36 dBm and 31 % at the same compression level, respectively.

Table 4.7 summarizes all of the relevant small-signal and large-signal measurement results of both HPA MMICs for comparison and evaluation of the use of source-connected shields. The expected increase in the small-signal gain performance of the HPA using HEMT devices with shields could be obtained as can be seen in Figure 4.12(a), but without the expected lower bandwidth. Thus, the bandwidth limitation accounts for broader bandwidths than that shown here. In both cases, there is almost no change due to the use of shields.

Nevertheless, the HPA Cesar V2 offers an improved small-signal matching capability and a frequency shift of about 0.5 GHz towards higher frequencies. In addition, a 2 dB higher $P_{\text{out}}$ could be obtained with the HPA Cesar V2 almost independent of the input power level

Table 4.7: Comparison of the small-signal and large-signal measurement results of the MMICs Cesar V1 and Cesar V2 in CW-mode and biased at $V_{\text{DS}}$= 28 V and $V_{\text{GS}}$= -2.6 V.

| **Main features** | **HPA Cesar V1** "gate without shield" | **HPA Cesar V2** "gate with shield" | **Evaluation** relating to the HPA Cesar V2 |
|---|---|---|---|
| Small-signal gain | $\geq$ 20 dB within 5.2 - 9.5 GHz | $\geq$ 20 dB within 5.8 - 10.2 GHz | improvement towards higher frequencies |
| **at $P_{\text{in}}$= 20 dBm and $f_0$= 9 GHz** | | | |
| $P_{\text{out}}$ | 36 dBm | 38 dBm | increase of 2 dB |
| PAE | 31 % | 37 % | 6 % enhancement |
| power gain | 20 dB | 19 dB | about the same |

and a significant increase in PAE up to an input power level of 23 dBm due to the benefit of field shaping. The expected benefits of the use of shields at higher operating voltages are shown in [57].

## 4.2 Impact of HEMT Structure and Layout on HPA Performance

A first PAE increase could be observed due to the modification of the gate module, i.e. the introduction of field plates. Furthermore, the HEMT structure and layout also have a strong impact on the HPA performance. This section deals with two aspects concerning the HEMT structure and layout, the impact of a change in the epitaxial structure, i.e. the aluminum (Al) concentration of the AlGaN layer, and the consequences of a variation in the HEMT layout, i.e. a reduction of the gate-to-gate pitch.

### 4.2.1 Impact of the Epitaxial Structure

Within the last section, the RF-performance of the HPA Reserva could be increased by the implementation of gate-connected field plates and the T-gate structures. A further technology progress includes the adaption of the Al content of the AlGaN layer with respect to the HEMT performance. Hereby, a reduction of the Al content of the $Al_xGa_{1-x}N/GaN$ HEMT device is a promising factor to improve the PAE behavior of the HEMT device and consequently that of the HPA design.

#### Influence of the Aluminum Mole Fraction on the HEMT Device

Within this work, the Aluminum concentrations in the barrier of the $Al_xGa_{1-x}N/GaN$ HEMT devices were reduced from $x = 30\,\%$ to $22\,\%$. Subsequently, the change in the device and HPA performance was analyzed.

According to the literature, a reduced Al content leads to a reduced $I_{max}$, lower leakage current, and less DC-to-RF dispersion. The lower gate leakage current results from a reduced electric field caused by a smaller polarization charge at the AlGaN/GaN interface [77]. Based on the results from Chapter 2.3, the above mentioned factors result in a PAE improvement. In addition, the reduced $I_D$ improves the reliability of the device thus improving the lifetime.

Nevertheless, these positive effects of the Aluminum-content reduction come at the expense of a lower broadband matching capability due to the decrease in $I_D$ and the increase in $C_{gd}$. Another effect of a change in the Al concentration is that the pinch-off voltage increases, i.e. tends towards more positive values, with decreasing Al content. Table 4.8 compares the intrinsic parameters of two T-gate AlGaN/GaN HEMT devices with 30 % and 22 % aluminum concentration in the barrier. According to the Table, the Al content has almost no impact on the capacitances besides a lower feedback capacitance $C_{gd}$ leading to an improved MSG/MAG behavior and accordingly a higher PAE. Nevertheless, the feedback loss $R_{gd}$ is higher for the device with the lower Al concentration.

Table 4.8: Comparison of the calculated intrinsic parameters of 1 mm AlGaN/GaN HEMT with different Al concentrations in the barrier measured at 30 V and $I_{DS}$ = 100 ... 150 mA.

| **Parameter** | **30 % Al** | **22 % Al** | **Evaluation** relating to HEMT with 22 % Al |
|---|---|---|---|
| $C_{gd}$ | 0.13 pF/mm | 0.1 pF/mm | decrease of 23 % ⇒ increase in MSG/MAG ⇒ increase in PAE |
| $C_{gs}$ | 0.92 pF/mm | 0.95 pF/mm | almost no change |
| $C_{ds}$ | 0.2 pF/mm | 0.21 pF/mm | almost no change |
| $g_m$ | ≈ 210 mS/mm | ≈ 237 mS/mm | small increase |
| $G_{ds}$ | 9.7 mS/mm | 5.4 mS/mm | decrease of 44 % |
| $R_{gd}$ | 10.3 Ωmm | 13.7 Ωmm | increase of 33 % |

#### HPA Performance dependent on the Aluminum Content in the AlGaN Layer

Again, the HPA Reserva was taken as an example to illustrate the changes in the small-signal and RF performance with decreasing Al content. The results shown in the following diagrams depict the small- and large-signal measurements of the HPA Reserva processed

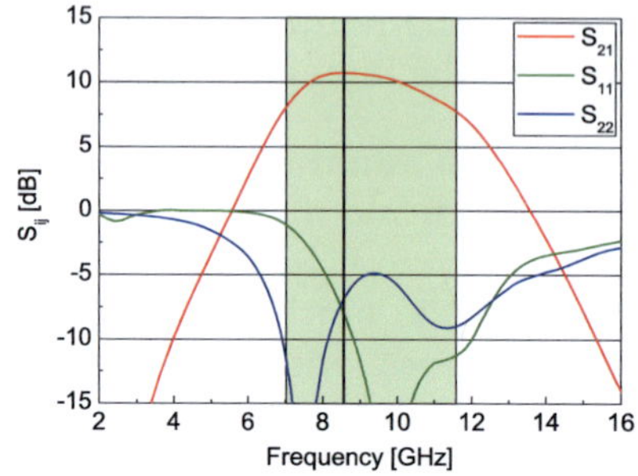


(a) CW-S-parameters at $V_{DS}$=28 V

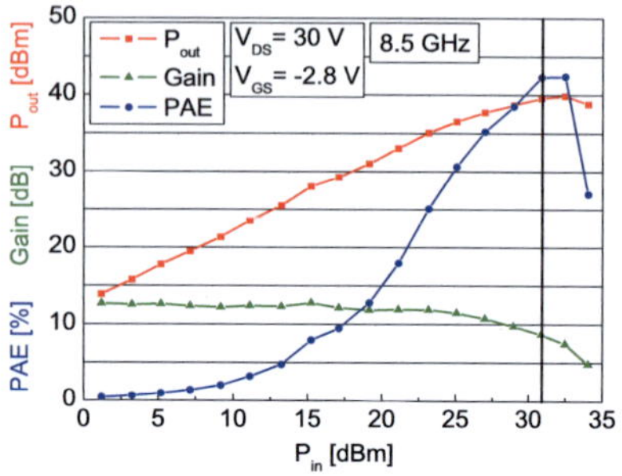


(b) Pulsed-RF performance at $V_{DS}$=30 V

Figure 4.13: Measured small-signal and large-signal performance of the power amplifier Reserva III (22 % Al content).

with the T-gate technology and an Al content of 22 %.

Figure 4.13(a) shows the measured CW-S-parameters of the MMIC Reserva biased at $V_{DS}$=28 V and $I_{DS}$=125 mA. A maximum small-signal gain of nearly 11 dB was obtained at a frequency of 8.5 GHz as marked in the figure. In addition, the corresponding input and output reflection coefficients around this frequency are better than -5 dB and the reverse isolation $S_{12}$, which is not shown in the figure, is less than -17 dB for the entire frequency range. In addition to the frequency of maximum gain, the 3 dB bandwidth from 7 up to 11.6 GHz is highlighted in green.

Figure 4.13(b) presents the pulsed-RF measurements of the HPA Reserva at the maximum small-signal gain frequency at 8.5 GHz and at a biasing of $V_{DS}$=30 V. A maximum PAE of 43 % was reached for a relatively high input power level of 31 dBm. The output power and associated power gain at the same input power level yield 39.6 dBm and 9 dB of gain with a compression of 3.5 dB.

**Discussion**

The measurement results obtained with the HPA Reserva III processed with the reduced Al content of 22 % deliver significant further performance improvements, especially for the large-signal case. Table 4.9 gives a comparison between the measurement results of the HPA Reserva II processed with a T-gate and an Al content of 30 % (see last section) and the measured data of the HPA Reserva III processed with a T-gate as well, but with a reduced Al content of 22 % at an input power level of 29 dBm. In addition, the optimum measurement results of the HPA Reserva III at $P_{in} = 31$ dBm are shown.

Table 4.9: Influence of the aluminum content on the HPA Reserva processed with 30 % Al content (HPA Reserva II) and 22 % Al content (HPA Reserva III) at $V_{DS}$= 30 V.

| **Main features** | **HPA Reserva II** "30 % Al content" | **HPA Reserva III** "22 % Al content" | | **Evaluation** relating to Reserva III |
|---|---|---|---|---|
| **LS results** | **at $f_0$= 8 GHz** | **at $f_0$= 8.5 GHz** | | |
| $P_{in}$ | 29 dBm | 29 dBm | 31 dBm | |
| $P_{out}$ | 38.1 dBm ≙ 6.5 W | 38.6 dBm ≙ 7.2 W | 39.6 dBm ≙ 9.1 W | + 0.5 dB |
| PAE | 25 % | 38.5 % | 42.3 % | factor of > 1.5 |
| power gain | 8.5 dB | 9.6 dB | 8.7 dB | + 1 dB |
| gain comp. | 3.5 dB | ≈ 3 dB | ≈ 4 dB | about the same |
| $P_{out,sat}$ | > 38.1 dBm | 39.8 dBm | | |
| linear gain | 12 dB | 12.6 dB | | + 0.6 dB due to higher $C_{gd}$ |
| $V_{GS}$ | -4 V | -2.8 V | | $V_{pinch\text{-}off}$ increases with decreasing Al content |

By comparing both amplifiers at the same input power level, a 0.5 dB increase in $P_{out}$ and an enhanced small-signal and large-signal gain performance of about 0.5 - 1 dB was obtained due to the higher gate-drain capacitance. Though, the resulting gain compression is nearly the same.

The reduction of the Al content in the AlGaN layer results in a lower DC-to-RF dispersion. In addition to the reduced Al content, the HPA Reserva III was processed with a stabilized T-gate technology, while HPA Reserva II was processed on the first run with the new T-gate technology. As a consequence of both factors, the PAE performance becomes significantly improved by a factor of 1.5 at the same input power levels and even more when comparing the maximum values (25 % and 42.3 %). The high PAE enhancement provides evidence that the influence of possible leakage currents and DC-to-

RF dispersion together with a stabilized T-gate process has a drastic impact on the PAE behavior of the HEMT device and consequently of the HPA design. Furthermore, due to the reduced $I_{max}$, the devices withstand higher input power levels. Therefore, the amplifier can be driven into saturation without difficulty. The saturated output power yields 39.8 dBm.

### 4.2.2 Impact of the Gate-to-Gate Pitch

In a further analysis the MMIC Cesar V2 from the last section was compared to the HPA Cesar V3 also using AlGaN/GaN-HEMTs with T-gates and source-connected shields, but with reduced gate-to-gate pitch of the transistors within the PA-stage and designed based on the same design goals as specified in Section 4.1.3.

The reduction of the gate-to-gate pitch ($l_{gg}$) of the HEMT device leads to a smaller device structure with reduced parasitics as can be seen in Table 4.10. Due to the strong decrease in $C_{gd}$, the small-signal gain increases resulting in an improved PAE behavior. Otherwise, the reduced gate-to-gate pitch causes higher thermal heating effects that cause an increased gate leakage current.

Table 4.10: Comparison of the calculated intrinsic parameters of 1 mm AlGaN/GaN HEMTs with different gate-to-gate pitch measured at $V_{DS} = 30$ V and $I_{DS} = 100 \ldots 150$ mA.

| **Parameter** | $l_{gg}$ **=50 μm** | $l_{gg}$ **=20 μm** | **Evaluation** relating to HEMT with $l_{gg}$ =20 μm |
|---|---|---|---|
| $C_{gd}$ | 0.07 pF/mm | 0.04 pF/mm | decrease of 43 %<br>⇒ increase in MSG/MAG<br>⇒ increase in PAE |
| $C_{gs}$ | 1.12 pF/mm | 0.99 pF/mm | decrease of about 10 % |
| $C_{ds}$ | 0.21 pF/mm | 0.19 pF/mm | decrease of about 10 % |
| $g_m$ | ≈ 237 mS/mm | ≈ 170 mS/mm | decrease of 30 % |
| $G_{ds}$ | 8 mS/mm | 8 mS/mm | unchanged |
| $R_{gd}$ | 19.3 Ωmm | 3 Ωmm | strong decrease of about 80 % |

### Design Configuration and Layout

The HPA Cesar V3 was designed based on the power amplifier Cesar V2. The only difference between the two variants consists in the HEMT devices in the PA-stage. While the transistors of the MMIC Cesar V2 has a source-to-gate spacing of 1 $\mu$m and a gate-to-drain spacing of 3.5 $\mu$m with a gate-to-gate pitch of 50 $\mu$m, the shrinked FET version used in the HPA Cesar V3 has a reduced gate-to-gate pitch of only 20 $\mu$m with 0.7 $\mu$m for the source-to-gate spacing and 2 $\mu$m for the gate-to-drain spacing. Thus, the relative source-to-gate and gate-to-drain spacings stay the same, while only the gate-to-gate pitch is reduced by a factor of 2.5. In the first stage, the same transistor is used as in the HPA Cesar V2.

Since there was no transistor model available for a FET with reduced electrical spacings, in this first-step analysis the same matching networks as for the HPA Cesar V2 were used for the design Cesar V3. Small-signal simulations of the HPA Cesar V3 with measured S-parameters of the shrinked device led to similar but slightly improved results compared with the HPA Cesar V2.

Figure 4.14 shows the chip image of the dual-stage HPA Cesar V3 with a chip size of $3.5 \times 2\,\text{mm}^2$. In the figure the PA-stage transistors are marked in terms of a red ellipse to highlight the shrinked FET version.

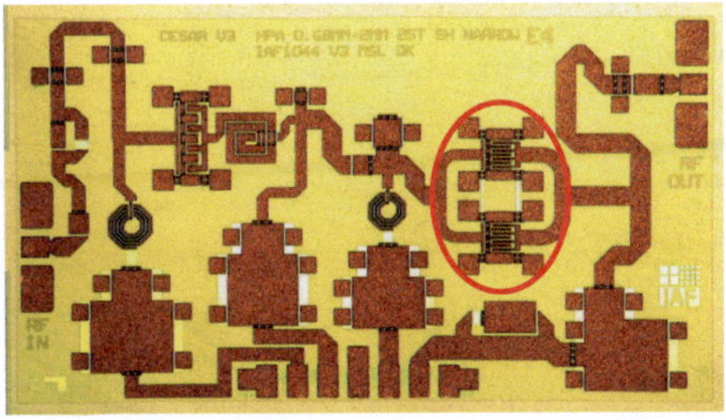


Figure 4.14: Photograph of the dual-stage HPA Cesar V3 with the shrinked FET version in the PA-stage marked. The chip size is $3.5 \times 2\,\text{mm}^2$.

### HPA Performance Dependence on the Gate-to-Gate Pitch

As well as for all previous comparisons, the measurement results shown below were from the same wafer with an even further reduced Al content of 18 % for improved output power and PAE values (see Section 4.2.1). Once again, the small-signal and large-signal measurements were taken in CW-operation at the same bias condition ($V_{DS}$= 28 V and $I_{DS}$= 100 mA) as the Cesar variants from the last section.

Figure 4.15 gives a comparison of the measured CW-S-parameters of the HPAs Cesar V2 (line without symbols) and Cesar V3 (line with symbols). The shrinking of the device structure leads to an increase in the small-signal gain of approximately 1-1.5 dB. The green area indicates the -3 dB frequency range of the HPA Cesar V3 from approximately 6.3-9.9 GHz with a small-signal gain beyond 25 dB. For both designs, the input and output matching $S_{11}$ and $S_{22}$ are in the range of -5 to -15 dB within the target band.

The comparison of both designs at the same bias condition shows that the shrinking of the device structure yields the expected increase of the small-signal gain of about 1-1.5 dB. Additionally, the input matching of both designs stays the same since there is no difference between both input matching networks and first-stage transistors. The

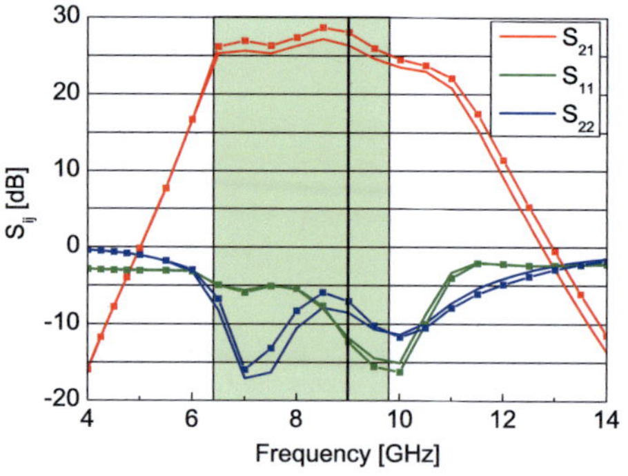


Figure 4.15: Comparison of the measured CW-S-parameters of the HPAs Cesar V2 (line without symbols) and Cesar V3 (line with symbols). Bias point: $V_{DS}$= 28 V and $I_{DS}$= 100 mA.

output matching of the HPA Cesar V3 turns out to be inferior compared to that of the HPA Cesar V2 because of the modified optimum load impedance $\Gamma_{L,opt}$ of the shrinked FET cells within the PA-stage combined with the output matching network of the Cesar V2 design optimized for $\Gamma_{L,opt}$ of the original HEMT device.

The large-signal performance of both Cesar variants is shown in Figure 4.16. The figure on the left gives a comparison of CW-power-sweeps at 9 GHz of the HPAs Cesar V2 and Cesar V3, while the right figure presents the corresponding frequency sweeps at $P_{in}$= 20 dBm, i.e. at the required 2 dB gain-compression level.

At a frequency of 9 GHz, the HPA Cesar V3 delivers a constant gain and output-power improvement of about 1 dB compared with the MMIC Cesar V2. Furthermore, the HPA Cesar V3 offers a PAE of 43 % at 2 dB compression, while the MMIC Cesar V2 delivers a comparatively lower PAE of 38 % at the same compression level.

In Figure 4.16(b) the frequency range from 8 up to 9.8 GHz in which both MMICs exhibit PAE values beyond 30 % is highlighted in green. The HPA Cesar V3 primarily features a PAE improvement of up to 5 % especially for the upper target band. In addition, the output power and associated gain are enhanced from approximately 8.5 GHz on. Thus, the HPA version using shrinked FET cells in the PA-stage delivers an overall improvement concerning the small-signal and large-signal performance. Thereby, the additional heating of the

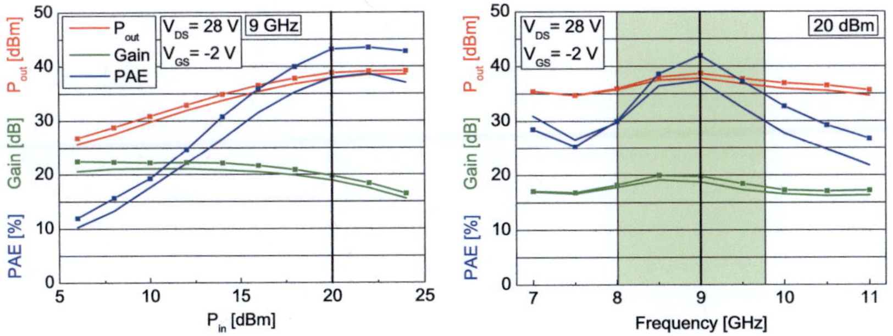


(a) Power sweeps at 9 GHz (b) Frequency sweeps at $P_{in}$= 20 dBm

Figure 4.16: Comparison of CW-power-measurements of the HPA MMICs Cesar V2 (line without symbols) and Cesar V3 (line with symbols). Bias point: $V_{DS}$= 28 V and $V_{GS}$= -2 V.

shrinked FET cells has to be considered and analyzed.

Table 4.11 summarizes the most important large-signal measurement results of the HPAs Cesar V2 and Cesar V3 at the same biasing conditions for good comparison. In addition, both HPAs offer HEMT devices with 18 % Al content in the AlGaN layer.

Table 4.11: Comparison of the CW-power measurement results of the MMICs Cesar V2 ($l_{gg}$ =50 $\mu$m) and Cesar V3 ($l_{gg}$ =20 $\mu$m) operated at 9 GHz and $P_{in}$= 20 dBm. Bias point: $V_{DS}$= 28 V and $V_{GS}$= -2 V.

| Main features | Cesar V2 $l_{gg}$ =50 $\mu$m | Cesar V3 $l_{gg}$ =20 $\mu$m | Evaluation relating to HPA Cesar V3 |
|---|---|---|---|
| $P_{out}$ | 38 dBm | 39 dBm | increase of 1 dB |
| PAE | 38 % | 43 % | factor of 1.13 in PAE |
| power gain | 19 dB | 20 dB | increase of 1 dB |

According to the LS-measurement results at 9 GHz, shown in Figure 4.16(a), the use of shrinked FET cells in the PA-stage delivers an improvement of the output power and power gain of approximately 1 dB nearly independent of the input power level, but frequency dependent according to the results of Figure 4.16(b). Based on the frequency-sweep measurement results, the shrinking of the gate-to-gate pitch delivers a large-signal performance improvement only for "higher" frequencies, i.e. in this case for frequencies above 8 GHz.

Consequently, the PAE of the HPA Cesar V3 could also be increased with higher enhancement effects towards larger input power levels and higher frequencies.

The chapter covered different technology aspects that have a strong impact on the HPA performance, especially the power-added efficiency. The differences between the technology modifications were analyzed by means of various corresponding narrow-bandwidth HPA designs processed with each technology. PAE improvements from 22.5 % up to over 40 % could be obtained with the combination of different technology modifications.

All further HPA designs within this work, which are discussed in detail in Chapter 5 and Chapter 6, use T-gate HEMT devices without source-connected shields, an aluminum concentration of the AlGaN layer of 22 %, and a standard gate-to-gate pitch of 50 $\mu$m.

# Chapter 5

# Balanced HPA with GaN-specific Impedance Levels

This chapter describes the first balanced AlGaN/GaN HEMT HPA MMIC demonstrator for X-band that is fabricated in microstrip technology. A few balanced MMIC amplifier designs at X-band frequencies exist in different technologies as described in [7], [51], [62], but no balanced microstrip HPA in GaN technology. Nevertheless, the study of such a power amplifier is promising with respect to the expected benefits of this concept using GaN technology compared to existing solutions with other technologies such as GaAs.

The main advantage of a GaN based balanced approach is the high load impedance level of GaN that be can used effectively in the balanced design as described in this chapter together with all the other principal advantages of the AlGaN/GaN material system (see Chapter 1) as well as the benefits of the balanced approach itself that is discussed in Section 5.2. For example, the high load impedance of a single-ended HPA with AlGaN/GaN HEMT technology leads to a poor output matching $S_{22}$ capability. Here, the main task of designing a balanced amplifier is the realization of a superior matching performance over a broader bandwidth together with a good stability.

However, a major problem of the realization of a balanced HPA MMIC in a 50 Ω environment based on SiC or GaAs substrates with an $\varepsilon_r$ between 9 and 13 is the design layout of a 50 Ω Lange coupler. This coupler features very narrow line widths and gaps that are too small for fabrication. In addition, the total line width of a 50 Ω coupler would not be able to ensure a sufficient power and current handling capability. To overcome this problem, the use of low-impedance microstrip Lange couplers on the input and output side is investigated. On the one hand, with the low-impedance, the critical line width and gap dimensions can be avoided and the resulting wider total line width enables the required current loading for RF output powers beyond 10 W. On the other hand, for the output coupler impedance, the optimum load impedance level of the GaN HEMT HPA in the 10-20 W power regime of approximately 25 Ω can be chosen to be able to simplify the impedance transformation networks between the PA modules (see below) and the Lange coupler. This would lead to a reduction of the required MMIC chip size as well as to an improvement of the HPA performance, i.e. to a simple and broadband matching circuit.

## 5.1 Principal Aspects of a Balanced Amplifier

The balanced amplifier, introduced by Eisele [32], basically consists of two identical amplifiers PA 1 and PA 2 as well as two quadrature (90 °) 3 dB couplers as can be seen in Figure 5.1. In the balanced topology, both amplifiers are fed equally with a phase offset of 90 ° from an input power splitter [136]. Any mismatch reflections at the amplifier inputs pass back through the couplers and appear 180 ° antiphase at the coupler input and therefore become canceled out or absorbed in the termination resistor $Z_0$. At the output, the power of the amplifiers is recombined in phase, and unwanted reflection signals are canceled out or absorbed in the termination resistor $Z_0$ just as for the amplifier input [1]. As a result, the matching at both the input and output of a balanced amplifier is excellent regardless of the reflection nature of the individual single-ended amplifier.

Independent of the 90 ° hybrid coupler type, the correlation of the small-signal scattering parameters of Port 1 and Port 2 of the entire balanced amplifier are given in terms of the small-signal S-parameters of the two individual single-ended amplifiers [117].

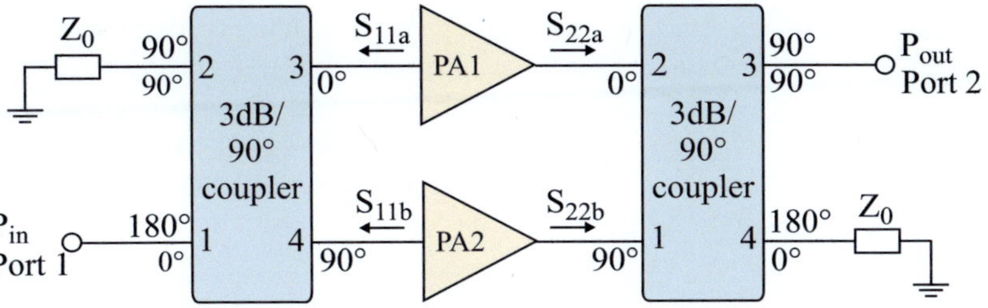


Figure 5.1: Illustration of a balanced amplifier design configuration.

$$|S_{11}| = \frac{1}{2}\,|S_{11a} - S_{11b}|, \qquad |S_{21}| = \frac{1}{2}\,|S_{21a} + S_{21b}| \tag{5.1}$$

$$|S_{12}| = \frac{1}{2}\,|S_{12a} + S_{12b}|, \qquad |S_{22}| = \frac{1}{2}\,|S_{22a} - S_{22b}| \tag{5.2}$$

According to Equations 5.1 and 5.2, the input and output reflection parameters of the balanced amplifier are zero regardless of the reflection behavior of the single-ended PAs as long as PA 1 and PA 2 are identical. This leads to a very low input and output small-signal VSWR of the balanced amplifier, independent of the VSWR of the single-ended amplifiers. In addition, the transducer gain of the balanced amplifier results in

$$G_{\mathrm{T}} \;=\; 0.25\,|S_{21a} + S_{21b}|^2 \;=\; |S_{21a}|^2 \tag{5.3}$$

which is identical to the gain of the single-ended amplifier in the ideal case. However, the saturated output power of the balanced amplifier is twice the output power of the single-ended PA. Furthermore, the S-parameter matrix of the quadrature (90°) 3 dB coupler is

$$S_{\mathrm{Coupler}} = -\frac{1}{\sqrt{2}}\begin{pmatrix} 0 & 0 & 1 & j \\ 0 & 0 & j & 1 \\ 1 & j & 0 & 0 \\ j & 1 & 0 & 0 \end{pmatrix}. \tag{5.4}$$

However, in real HPA designs for large-signal (LS) applications, the power amplifier usually has "large-signal" output matching which ensures a proper loadline setting at the intrinsic transistor output current source instead of an accurate small-signal conjugate match because in the general case of an arbitrary cell size both conditions are mutually exclusive. To analyze the large-signal behavior of the balanced amplifier the wave equations at the output coupler are determined according to Figure 5.2.

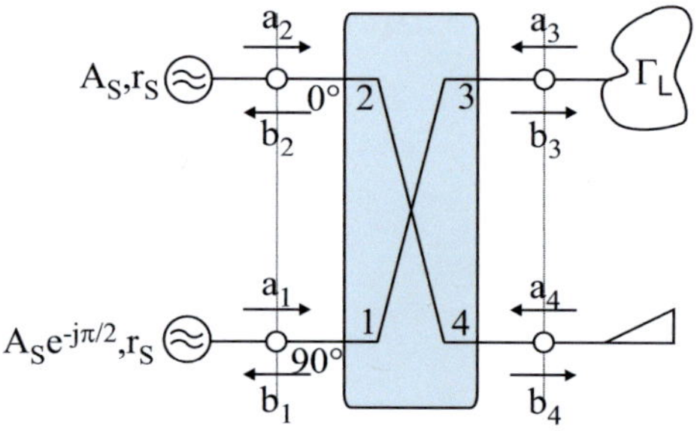


Figure 5.2: Lange coupler at the output side of a balanced amplifier design with incoming $a_i$ and outgoing $b_i$ waves.

Assuming that PA 1 and PA 2 are identical, the magnitudes of the amplitudes $|A_s|$ of the incoming waves $a_1$ and $a_2$ amount to the same value, but with 90° phase difference. This leads to the following components of $\vec{a}$.

$$\begin{aligned} a_1 &= jA_s + r_s b_1 \\ a_2 &= A_s + r_s b_2 \\ a_3 &= \Gamma_L b_3 \\ a_4 &= 0\,. \end{aligned} \tag{5.5}$$

With $[b] = \left[S_{\text{Coupler}}\right] \cdot [a]$ and some conversions the components of

$[b]$ result in

$$
\begin{aligned}
b_1 &= -\frac{1}{\sqrt{2}}\Gamma_L b_3 && = j\Gamma_L A_s \\
b_2 &= -\frac{j}{\sqrt{2}}\Gamma_L b_3 && = -\Gamma_L A_s \\
b_3 &= -\frac{1}{\sqrt{2}}(jA_s + r_s b_1) - \frac{j}{\sqrt{2}}(A_s + r_s b_2) && = -j\sqrt{2}A_s \\
b_4 &= -\frac{j}{\sqrt{2}}(jA_s + r_s b_1) - \frac{1}{\sqrt{2}}(A_s + r_s b_2) && = \sqrt{2}r_s\Gamma_L A_s \,. \quad (5.6)
\end{aligned}
$$

On closer examination of Port 4, the power that is delivered to the termination resistor is

$$P_4 = \frac{1}{2}\left|b_4\right|^2 = \left|r_s\Gamma_L A_s\right|^2 \,. \quad (5.7)$$

In case of a perfectly matched PA module ($r_s = 0$) no power is delivered to the termination resistor, and with increasing reflection behavior of the PAs, the power at the termination resistor also increases. Thus, the higher the reflection coefficient $r_s$ for a given load mismatch $\Gamma_L$, the more of the reflected part of the available power of the two single-ended amplifiers is absorbed by the termination resistor instead of being reflected back into the PA transistors, indicating a good protection for the PA-FETs.

Based on Equations 5.6, the reflection coefficients at Port 1 ($b_1/a_1$) and Port 2 ($b_2/a_2$) can be calculated as follows:

$$
\begin{aligned}
\frac{b_1}{a_1} &= \frac{j\Gamma_L A_s}{jA_s + jr_s\Gamma_L A_s} &= \frac{\Gamma_L}{1 + r_s\Gamma_L} \\
\frac{b_2}{a_2} &= \frac{-\Gamma_L A_s}{A_s - r_s\Gamma_L A_s} &= \frac{-\Gamma_L}{1 - r_s\Gamma_L} \,.
\end{aligned} \quad (5.8)
$$

As can be seen in Equation 5.8, the reflection coefficients at Port 1 and Port 2 are not identical. Even if the single-ended PAs are perfectly matched, i.e. $r_s = 0$, both PAs directly "see" the load reflection coefficient $\Gamma_L$, but with 180° phase difference due to the coupler. Consequently, the balanced amplifier delivers no improvement concerning the large-signal reflection behavior compared with the single-ended PA.

## 5.2 Advantages Over Single-Ended Amplifier

Even though the large-signal reflection behavior of the balanced amplifier is not superior to that of the single-ended PA, the balanced amplifier features quite a lot of advantages over a single-ended amplifier according to [40]:

- reduction of input and output small-signal reflection coefficients
- superior small-signal matching bandwidth which is mainly limited by the coupler bandwidth
- almost doubled maximum possible output power due to the two parallel amplifiers, minus the coupler losses
- improved stability due to high degree of decoupling
- redundancy in hybrid designs: in case one of the PAs fails, the other one can still work so that the balanced amplifier will operate with reduced power gain (-6 dB)
- advantageous for application fields where strong output and / or input mismatch-es occur, e.g. high power measurement setups or X-band near-field radars.

However, a balanced amplifier is expensive due to the need for a large chip area compared with a single-ended amplifier.

## 5.3 Determination of Coupler Type

When designing a quadrature (90°) 3 dB balanced amplifier, one of the first questions addresses the coupler itself and what type of coupler would be the best for a specific application. In this work, the advantages and disadvantages of a Lange coupler are compared with that of a Branch-Line coupler, since both coupler types are realizable in microstrip line technology. Block diagrams of both coupler structures are illustrated in Figure 5.3. As can be seen in the figure, both networks are symmetrical.

The method of the Branch-Line coupler was first analyzed by Reed and Wheeler [91]. The advantages of this coupler are its simple structure and the easy 50 Ω matching. However, the Branch-Line coupler requires a large chip area because all lines are a quarter of the

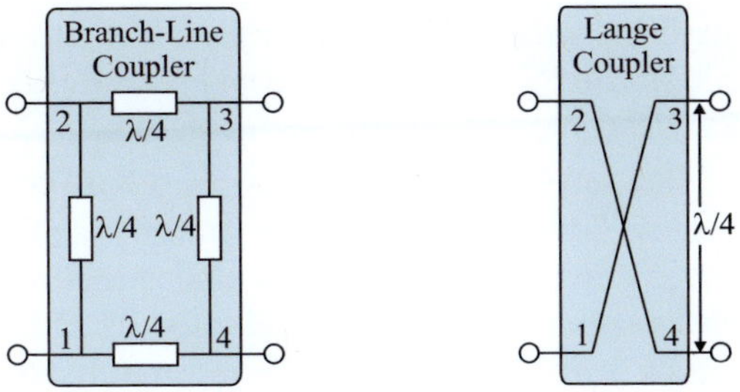


Figure 5.3: Comparison of the structure of a Branch-Line coupler and a Lange coupler.

wavelength long. At a frequency of $f_0 = 10\,\text{GHz}$, the wavelength of a microstrip line on SiC substrate with a relative dielectric constant of $\varepsilon_r = 9.7$ only depends on the ratio of the variable microstrip line width $w_c$ and the fixed height of the substrate ($h = 100\,\mu\text{m}$) leading to the well-known equation 5.9

$$\lambda/4 = \frac{1}{4} \cdot \frac{c_0}{f_0 \cdot \sqrt{\varepsilon_{r,\,eff}(w_c/h)}} \,. \tag{5.9}$$

With a typical $50\,\Omega$ microstrip line resulting in a line width of $w_c = 92\,\mu\text{m}$, the effective relative diel. constant amounts to $\varepsilon_{r,\,eff} = 6.5$ according to the nomogram of the effective relative dielectric constant $\varepsilon_{r,\,eff}$ as a function of the ratio $w_c/h$ as given e.g. in [117]. As a result, the quarter wavelength is about $\lambda/4 = 2940\,\mu\text{m}$. To reduce the size of this coupler, the resulting $\lambda/4$ lines can be decreased by the use of dual transmission lines as described in [114]. Nevertheless, the coupler still requires a large chip size. Besides that, a realization of a Branch-Line coupler at lower impedance levels, which would be advantageous for simplified matching of the HPA modules (see below), would require an even larger chip area and is not feasible for this reason. In addition, the obtainable relative bandwidth with this coupler type only amounts about $BW_{rel}$= 10-20 % according to [45].

In contrast, the Lange coupler with its smaller structure requires less chip size. Furthermore, a wider frequency range can be obtained with this coupler. Its mechanism is originally explained in full length in [61]. However, Lange couplers with a nominal impedance of $50\,\Omega$ are not realizable on SiC substrate with a thickness of $100\,\mu\text{m}$ because of the resulting small line widths and low current capability as described in detail in Section 5.4.2. To overcome this problem, the design approach used here relies on low-impedance Lange couplers.

Balanced MMIC amplifier designs at X- and Ku-Band frequencies are summarized in Table 5.1. The table mainly contains balanced HPAs using Lange couplers, but there is also one amplifier containing Branch-Line couplers and one employing 0° hybrid Wilkinson couplers. As can be seen in the table, the obtainable frequency bandwidths of the balanced applications mainly depends on the coupler type. Only narrow bandwidths are realizable with Wilkinson and Branch-Line couplers compared with Lange couplers.

Due to the advantageous properties of the Lange coupler such as the required smaller chip size and the larger obtainable frequency bandwidth according to the table, this coupler type is used for the balanced amplifier design discussed in the following.

Table 5.1: Balanced MMICs at X- and Ku-Band found in literature.

| **Ref.** | **Technology + Chip Size** | **Coupler Type** | **Amplifiers used** | **Frequency + relative Bandwidth** | **Return Loss + + Gain** |
|---|---|---|---|---|---|
| [39] | ion-implanted GaAs MESFET not on one chip | Branch-Line coupler | 2-stage HPAs | 14-16 GHz $BW$= 13 % | not spec. G>13.7 dB |
| [7] | AlGaN/GaN HEMT (CPW) 25.4 $mm^2$ | Wilkinson coupler | 1-stage HPAs | 7-8.6 GHz $BW$= 20 % | $S_{ii}$<-10dB G> 5 dB |
| [62] | ion-implanted GaAs MESFET 16 $mm^2$ (CPW) | Lange coupler | 2-stage HPAs | 8-12 GHz $BW$= 40 % | $S_{ii}$<-20dB G> 13 dB |
| [107] | AlGaN/GaN HEMT (CPW) 15.3 $mm^2$ | Lange coupler | 2-stage LNAs | 3-16 GHz $BW$= 130 % | not spec. G= 20 dB |
| [68] | AlGaN/GaN HEMT (CPW) 20.6 $mm^2$ | Lange coupler | NDPAs | 4-18 GHz $BW$= 130 % | not spec. G= 10 dB |
| [56], this work | AlGaN/GaN HEMT (MSL) 12 $mm^2$ | Lange coupler | 1-stage HPAs | 7-11 GHz $BW$= 40 % | $S_{ii}$<-10dB G> 8 dB |

## 5.4 Balanced-HPA Implementation

A balanced microstrip AlGaN/GaN HEMT HPA MMIC was realized with a modular circuit concept for the purpose of good testability of the single amplifier modules, in this case single-ended 50 Ω terminated PAs, named Reserva, and "low-impedance" Lange couplers.

The block configuration of the resulting balanced HPA MMIC, named Crianza, is shown in Figure 5.4, illustrating the modular circuit concept using two identical single-stage amplifiers (PA 1 and PA 2), impedance level transformers, and low-impedance Lange couplers. The Lange couplers are located at the input and output ports of the PA modules to improve the small-signal return losses and to achieve simple and easy matching of the balanced design.

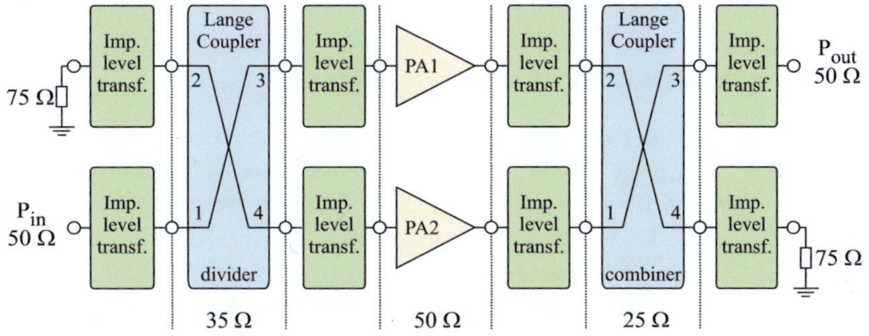


Figure 5.4: Block diagram of the balanced HPA MMIC Crianza.

The impedance level transformers between the power amplifiers and couplers are necessary because of the modular concept comprising different impedance levels of the power amplifiers and Lange couplers. In this circuit design, the impedance matching networks are simple line-capacitor circuits operating in the frequency range from 8 to 10 GHz with good performance. Due to the approximately $\lambda/8$ line lengths of the transformers, these line-capacitor networks just fit between coupler and PA 1/2 as well as between coupler and input and output ports of the balanced HPA and the termination resistors. For the design of the entire balanced amplifier different design steps are mandatory:

1. PA module terminated to a 50 $\Omega$ environment
2. GaN specific impedance-level selection of Lange couplers
3. Realization of balanced HPA

### 5.4.1 PA Module

In the first design step the design of the single-ended power amplifier Reserva is defined. The PA module demonstrates a single-stage amplifier containing a transistor stage and an input and output matching network (IMN and OMN). The transistor stage consists of two parallel operated 1 mm TGW basic AlGaN/GaN-HEMT cells with an $8\times125\,\mu$m gate geometry as a best compromise between gate finger count and finger width to obtain optimum gain performance as described in Section 3.1.1. Consequently, the total gate-width is 2 mm. The HEMT devices are based on the T-gate technology without source-connected shields and a gate-to-gate pitch of 50 $\mu$m. The aluminum content in the barrier is 22 %.

Since the PA is dimensioned for maximum output power operation in the frequency range from 8 to 10 GHz, Load-Pull simulations of the $8\times125\,\mu$m FET cell are performed within the desired X-band frequency range to specify the optimum load impedance $\Gamma_L$ of the used HEMT cells with respect to maximum output power.

After the determination of the optimum load impedance $\Gamma_{L,Pout}$, both input and output matching networks are designed for a 50 $\Omega$ external load. The networks comprise simple narrowband line-capacitor structures for the required impedance level transformation and matching over the desired frequency bandwidth as well as bias feeder paths that are realized with microstrip lines and large bias decoupling capacitors at the end of the bias stubs to provide DC blocking. In addition, small series capacitors for DC decoupling are included close to the input and output of the input and output matching networks, respectively.

To realize external stability, a resistor is included in the gate bias path and to prevent any odd-mode oscillations (internal stability) the gate and drain connections of both FET cells are connected together through resistors and transmission lines. Further details of the determination of external and internal stability are described in Section 3.3.1.

Figure 5.5 displays the chip image of the single-stage HPA MMIC Reserva with a chip size of $2.75 \times 1.5\,\mathrm{mm}^2$. The layout also includes DC bias pads for direct on-wafer tests.

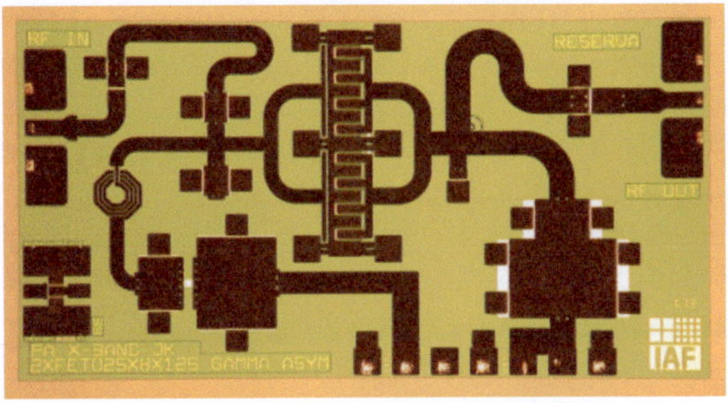


Figure 5.5: Photograph of the single-stage HPA Reserva. The chip size is $2.75 \times 1.5\,\mathrm{mm}^2$.

### 5.4.2 GaN-Specific Impedance-Level Selection of Lange Couplers

The second design step covers the choice of adapted impedance levels for the 3 dB Lange couplers at the input and output of the entire balanced amplifier as well as the structure and design of the couplers themselves. There are some limitations on the Lange coupler geometries. For example, a minimum line width $w_c$ and gap $s_c$ are indispensable for electroplated metal. In addition, the gap $s_c$ between the single lines strongly influences the coupling factor which therefore turns out to be sensitive to MMIC technology deviations, especially in galvanic metallization structures. Furthermore, the total coupler line width, calculated from the product out of the number of lines $N_{L,c}$ and the single line width, indicates the current carrying capability. Thereby, a change of the impedance level of the Lange coupler has a direct impact on its geometries. When lowering the impedance level, the RF current amplitude for a given power level increases inversely proportional to $\sqrt{Z_L}$, i.e. $P = 1/2\ Z_L \hat{I}^2$ or $\hat{I} \sim 1/\sqrt{Z_L}$. A low impedance level of the coupler is advantageous because the total conductor width increases disproportional with lowering $Z_L$ leading to an increased current capability.

Table 5.2 shows the influence of the impedance level $Z_L$ on 4-line and 6-line 3 dB Lange coupler geometries for a metal thickness of $t$=7 $\mu$m realized on SiC substrate with a thickness of $h$=100 $\mu$m and with an $\varepsilon_r$=9.7. All dimensions are calculated with Agilent ADS LineCalc centered to 10 GHz. Due to the requirement for high output power levels, a 6-line Lange coupler is chosen at the output to ensure high power and current handling capabilities. In contrast, on the input side, a 4-line coupler is sufficient. The expected width $w_c$ and gap $s_c$ for both coupler variants at an impedance of $Z_L$=50 $\Omega$ are displayed first. In both cases the width of the single line would be too small for fabrication. As a result, "low-impedance" microstrip couplers must be chosen for the realization on SiC substrate.

Table 5.2: Influence of the impedance level on the 3 dB Lange coupler geometries for $h$=100 $\mu$m and $\varepsilon_r$=9.7.

| **Coupler Type $N_{L,c}$** | **Impedance $Z_L/\Omega$** | **Width $w_c/\mu$m** | **Gap $s_c/\mu$m** | **Length $l_c/\mu$m** |
|---|---|---|---|---|
| 4-Line | 50 | 7.8 | 8.3 | 2741 |
| 4-Line | 35 | 25.1 | 7.8 | 2717 |
| 6-Line | 50 | 3.9 | 14.2 | 2550 |
| 6-Line | 25 | 29.3 | 13.3 | 2671 |

At first, the impedance level of the 6-line output Lange coupler was investigated. For a HPA with an expected output power beyond 20 W and consequently a high RF-current load, the impedance of the Lange coupler needs to be sufficiently lowered compared to 50 $\Omega$ so that the coupler provides sufficient power and current handling capability. In addition, a 2 mm PA-stage of an AlGaN/GaN HEMT HPA requires a load impedance of about 25 $\Omega$ at a drain voltage of 30-35 V (with $V_{knee} \approx 5$ V, $\hat{V} \approx 25$ V, $\hat{I} \approx 1$ A), combined with an appropriate $C_{out}$ compensation. As a result, a coupler that is adjusted to the optimum load impedance of 25 $\Omega$ of a typical GaN PA-stage enables a direct simple matching of the PA-module to the coupler. This direct matching is not exploited in this "first demonstrator" study but rather to provide a direct testability of the 50 $\Omega$ PA-module and the low-impedance coupler itself. Furthermore, a 25 $\Omega$ coupler ensures sufficient power and current handling capability and avoids critical microstrip line width and gap dimensions as can be seen in Table 5.2. Therefore, a 25 $\Omega$ 6-line Lange coupler is chosen at the output side.

In contrast, the load impedance level of a GaAs HEMT with a comparable output power level yields only 5-8 Ω. As a result, a comparable 5-8 Ω coupler on GaAs substrate with $\varepsilon_r = 12.9$ for direct matching with the PA module is not realizable. Therefore, when this balanced concept is applied to GaN, a simpler and broader matching circuit can be realized.

Regarding the input coupler, the power level at the input of the PA modules is about 2-4 W with an assumed large-signal gain of approximately 10...7 dB. Due to the resulting low current flow at the input side, in this case, a 4-line coupler delivers enough current handling capability with the positive side-effect of a lower coupler size that results in less required chip area. To provide a sufficient current handling capability on the coupling lines as well as to keep the minimum possible line width of the single lines in mind, the realized impedance of the input 4-line coupler is chosen to be 35 Ω.

To fit the balanced HPA design into a chip area with a width of 3 mm, the mechanical length $l_c$ of each coupler has to be smaller than 3 mm minus the upper and lower scribe lines of each 65 $\mu$m as well as the width of the connecting lines between the Lange couplers and the impedance level transformation networks. The resulting mechanical lengths of the calculated input and output couplers are 2717 $\mu$m and 2671 $\mu$m, respectively. This means that the 35/25 Ω couplers fit into the chip size without "folding" the lines. The layout of the output Lange coupler with crossover connections at both sides and in the center is shown in Figure 5.6.

The test structure of this Lange coupler was measured on-wafer

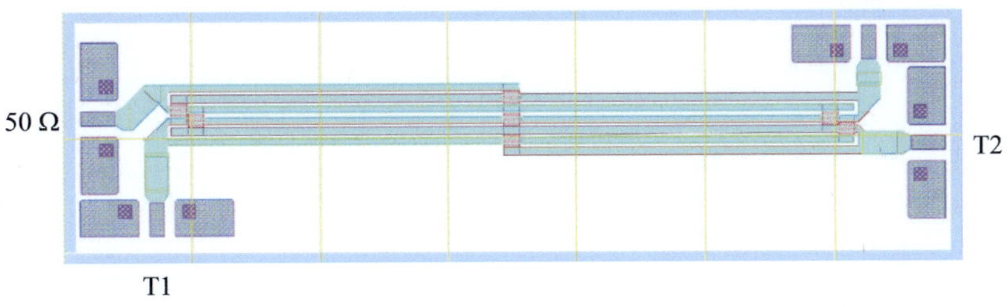


Figure 5.6: Layout of the microstrip 6-line 3 dB Lange coupler.

in a two-port 50 Ω environment with the two unused ports terminated by 50 Ω resistors. Next, the measured data was transformed into a 25 Ω four-port matrix and compared with the in ADS simulated coupler with a 25 Ω port impedance. Figure 5.7(a) gives a comparison of the simulated (solid line) and measured (dashed line) through connection ($S_{31}$) and coupling factor ($S_{32}$) of the output coupler. Simulated and measured data are in good agreement for lower frequencies up to the center frequency of 10 GHz as marked in the figure with a coupling factor of 3.5 dB ± 0.4 dB for both simulation and measurement from 8 to 10 GHz. However, for higher frequencies the measured S-parameters show a decreasing coupling effect which can be explained by the difference between the simulated and realized gap between the single coupler lines. Nevertheless, the symmetry performance of the coupler is still acceptable.

In addition, the simulated and measured isolation ($S_{21}$) is better than -20 dB up to a frequency of 10 GHz and the simulated and measured directivity at 10 GHz yields 22 dB and 17 dB, respectively.

Figure 5.7(b) shows the simulated and measured net insertion loss of the output coupler which amount to 0.23/0.29 dB and are in good

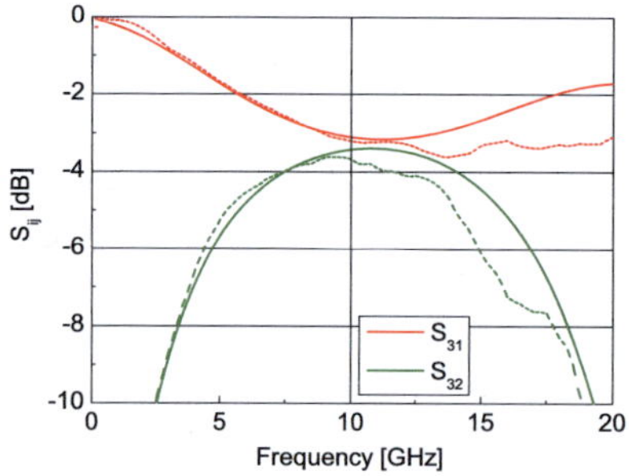


(a) Through connection $S_{31}$ and coupling factor $S_{32}$

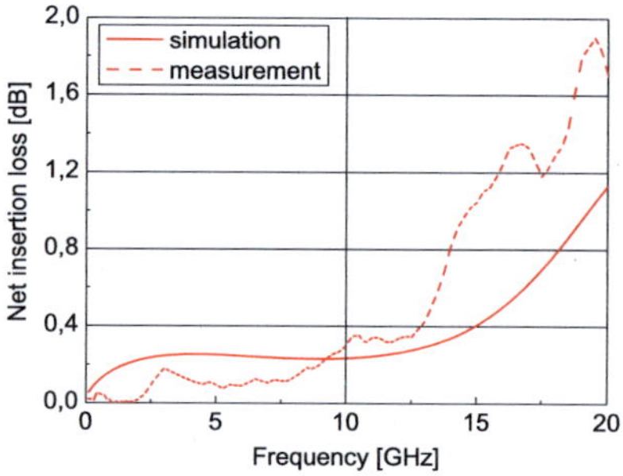


(b) Net insertion loss $L_{net}$

Figure 5.7: Comparison of the simulated (solid line) and measured (dashed line) through connection, coupling factor, and net insertion loss of the output 6-line microstrip Lange coupler referenced to 25 Ω.

agreement. Here, the net insertion loss is defined as

$$L_{\text{net}} = \frac{|S_{31}|^2 + |S_{32}|^2}{1 - |S_{11}|^2} \tag{5.10}$$

$$L_{\text{net}}\,[\text{dB}] = -20\log(\sqrt{L_{\text{net}}}\,)\,. \tag{5.11}$$

A low net insertion loss of the output Lange coupler is extremely important for optimum power transfer from the amplifier stages to the output load. This is achieved in spite of compact realization on a 100 $\mu$m thin SiC MMIC substrate with rather small line cross-section.

### 5.4.3 Realization of Balanced HPA

Finally, the Lange couplers and two of the PA modules (PA 1 and PA 2) are integrated to the balanced HPA Crianza with a total gate width of 4 mm according to the block configuration in Figure 5.4. The resulting design approach presents a first demonstrator for a balanced microstrip HPA MMIC using AlGaN/GaN HEMTs with the main task of the testability of the single amplifier modules.

Due to the fact that the PA modules and the couplers yield different impedance levels, impedance level transformation networks are necessary between the single-ended PAs, designed for 50 Ω external loads, and the Lange couplers, designed for 35 Ω at the input side and 25 Ω at the output side. In addition, impedance level transformers between the Lange couplers and the 50 Ω input and output of the balanced amplifier as well as between the couplers and the termination resistors are needed. The isolated ports of the Lange couplers are terminated by $Z_L$=75Ω resistors. On the one hand, this high load impedance level leads to a better current drive capability due to an increase of $Z_L$ by the factor 1.5 with a given maximum line width. On the other hand, the higher resistance leads to a slightly reduced bandwidth, because an additional matching network from 25/35 Ω up to 75 Ω is required.

The entire balanced HPA Crianza has a chip size of 4×3 mm$^2$. The chip photograph is shown in Figure 5.8.

In a final solution, the impedance level transformations between the PA modules and the couplers would be redundant leading to a simpler and more compact design that requires less MMIC chip area. In addition, the expected lower net insertion loss of the direct

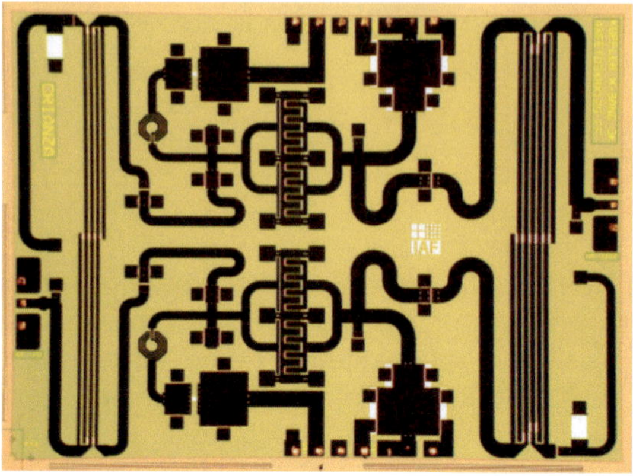

Figure 5.8: Photograph of the balanced single-stage HPA Crianza. The chip size is $4 \times 3\,\mathrm{mm}^2$.

impedance matching of the PA modules and the couplers would lead to an improved $S_{21}$-bandwidth performance of the entire HPA.

## 5.5 Balanced vs. Single-Ended HPA Performance

The simulated and measured CW-S-parameters of the amplifier module Reserva and the balanced HPA Crianza are plotted in Figure 5.9. For a reasonable comparison, both HPAs were measured on the same wafer and tested at the same biasing of $V_{DS}$= 28 V. Simulation and measurement reveal an excellent agreement over the whole frequency range for both designs.

A maximum small-signal gain of 10.7 dB was obtained for both the PA module Reserva and the balanced HPA Crianza at a frequency of 8.6 GHz and 8.8 GHz, respectively. The associated input and output reflection coefficients at the frequencies of best gain performance are $S_{11}$ = -8.3 dB and $S_{22}$ = -6.6 dB for the amplifier Reserva and for the balanced approach $S_{11}$ = -20.1 dB and $S_{22}$ = -23.6 dB. As expected, with the balanced design a strong improvement of input and output matching could be obtained. The same accounts for the broadband matching capabilities over the entire X-band frequency range.

While the input and output matching of the single-ended amplifier are about -5 dB from 8.5-13 GHz, both input and output matching of the balanced amplifier are strongly improved to a matching of better than -14 dB for the same frequency bandwidth.

Nevertheless, the HPA Reserva delivers a 3 dB bandwidth with a gain $\geq$ 8 dB from 7 up to 11.5 GHz, i.e. a relative bandwidth of $BW_{rel} = 45\,\%$, whereas the balanced HPA Crianza offers a slightly reduced corresponding 3 dB bandwidth with a gain $\geq$ 8 dB from 7.3 up to 11.1 GHz, i.e. a relative bandwidth of $BW_{rel} \approx 40\,\%$. In both graphics of Figure 5.9 the 3 dB frequency bandwidth is highlighted in green.

According to the theoretical assumption that the small-signal gain of the balanced HPA is equal to that of the used single-ended PAs as long as the PAs are identical (see Equation 5.3), both the balanced and the single-ended HPA show the same maximum small-signal gain. However, the balanced amplifier shows its maximum gain performance only over a very narrow bandwidth that corresponds to the optimum frequency range of the Lange couplers and the impedance level transformation networks that are optimized only for a frequency range from 8 up to 10 GHz. Thus, the reason for the reduced 3 dB bandwidth of the balanced topology mainly consists in the additional impedance transformations from 35 / 25 Ω to 50 Ω between the IMN / OMN and the couplers for the benefit of the measurability of the single modules.

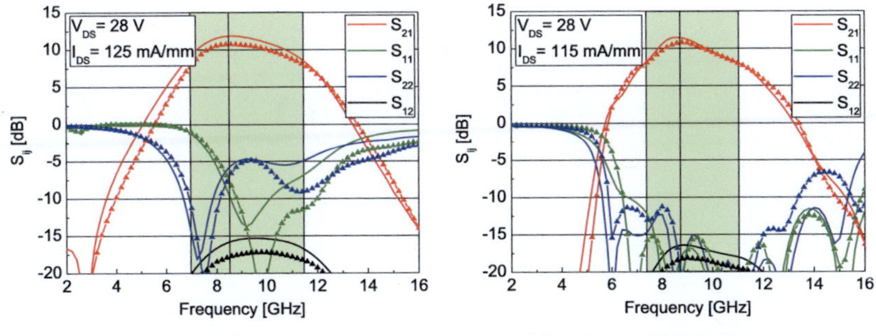


(a) Single-ended HPA Reserva (b) Balanced HPA Crianza

Figure 5.9: Simulated (line without symbols) and measured (line with symbols) CW-S-parameters of the HPAs Reserva and Crianza at $V_{DS}$= 28 V.

In contrast, the input and output matching of the balanced amplifier are better than -14 dB from 8.5 to 13 GHz which is a considerable advantage compared with the single-ended PA Reserva with an input and output matching of only better than -5 dB for the same frequency range. In addition, the measured reverse isolation $S_{12}$ is better than -17 dB for both HPAs at all frequencies.

Figure 5.10 presents the large-signal measurement results of both the single-ended amplifier Reserva and the balanced HPA Crianza. Output power, gain, and PAE measurements were performed in pulsed-RF mode with a 10 % duty cycle and 100 $\mu$s pulse width. The measured data are displayed for a frequency of 8.5 GHz at a drain voltage of $V_{DS}$= 25 V over an input power range up to 35 dBm. According to the small-signal measurement results both the balanced HPA Crianza and the PA Reserva show nearly the same linear gain performance at this frequency. The slightly reduced linear gain of the balanced approach is partly due to the small additional net insertion loss of the output coupler that is below 0.2 dB at 8.5 GHz as can be seen in Figure 5.7(b), and mainly caused by the different gate bias settings near pinch-off.

The single-ended amplifier Reserva delivers an output power of 39 dBm (7.9 W) at 3 dB gain compression with an associated PAE of 46 % and a power gain of 8 dB for 31 dBm input power. For a higher input power, a saturated output power of nearly 40 dBm is reached. This leads to a power density of nearly 4 W/mm at 3 dB gain com-

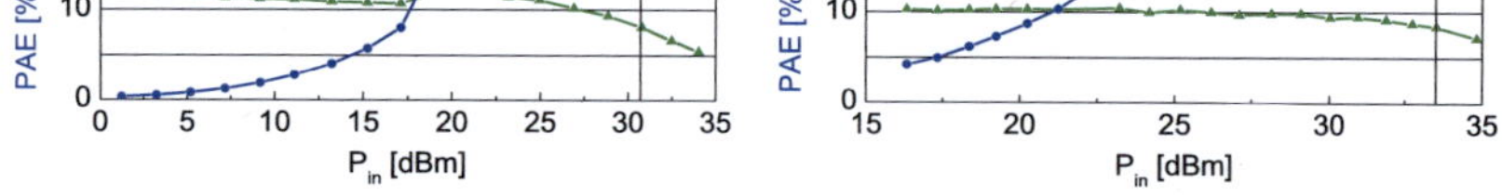


(a) Single-ended HPA Reserva (b) Balanced HPA Crianza

Figure 5.10: Pulsed-RF measurements of the HPAs Reserva and Crianza at $V_{DS}$= 25 V.

pression and up to 5 W/mm at saturation.

The maximum output power of the balanced amplifier Crianza is 3 dB higher compared with the PA module, i.e. 42 dBm (15.8 W), at about the same gain compression, but at a 2 dB higher input power level ($P_{in}$= 33 dBm). As a result, the power density of the balanced HPA Crianza reaches almost 4 W/mm at a 2 dB gain compression level which is comparable to that of the PA module. As expected, the associated power gain of 8 dB is equal to that of the PA Reserva. In addition, likewise the obtained PAE is nearly identical. There is only a small reduction from 47 % down to 44 % compared with the PA module.

The obtained measurement results show that a balanced microstrip MMIC HPA using AlGaN/GaN HEMTs delivers an advantageous circuit concept and a low-loss output (and input) matching network realization compared with the single-ended PA module. This is done on the basis of a skillful utilization of the high load impedance level of a GaN HEMT in the 10 to 20 W power regime, in contrast to the much lower impedance levels in, e.g., GaAs pHEMTs.

## 5.6 Conclusions

The developed balanced HPA Crianza delivers some important advantages over the single-ended PA Reserva. The most important small-signal and large-signal measurement results of both HPAs are summarized and compared in Table 5.3.

Based on the theoretical considerations given at the beginning of this chapter and due to the low net insertion loss of the output coupler at low X-band frequencies in the case of the balanced amplifier, it turns out that the small-signal maximum gain of both HPAs, the single-ended and balanced approach, is equal at about the same frequency of 8.6 and 8.8 GHz, respectively. As expected, the balanced HPA Crianza offers a strongly improved small-signal matching capability over the entire X-band frequency range. In addition, the balanced amplifier shows a very good stability because of the high degree of decoupling of the two amplifiers PA 1 and PA 2 and due to the fact that the balanced design does not cause additional possible ring-oscillations at the interconnection of both PA modules.

The relative 3 dB bandwidth $BW_{rel}$ with a gain performance $\geq$ 8 dB

Table 5.3: Comparison of the small-signal and large-signal measurement results of the single-ended HPA Reserva and the balanced HPA Crianza.

| **MMIC** | **HPA Reserva** | **HPA Crianza** | **Evaluation** relating to PA Crianza |
|---|---|---|---|
| Small-Signal S-Parameters performed at $V_{DS}$= 28 V | | | |
| $G_{max}$ | 10.7 dB at 8.6 GHz | 10.7 dB at 8.8 GHz | same $G_{max}$ at about the same frequency |
| $S_{ii}$ at $G_{max}$ | $S_{11}$= -8.3 dB<br>$S_{22}$= -6.6 dB | $S_{11}$= -20.1 dB<br>$S_{22}$= -23.6 dB | strong improvement of input and output matching |
| $S_{ii}$ within 8.5-13 GHz | $S_{ii} \leq$ -5 dB | $S_{ii} \leq$ -14 dB | strongly improved small-signal matching |
| $BW_{rel}$ with G$\geq$ 8 dB | 7-11.5 GHz<br>$BW_{rel}$ = 45 % | 7.3-11.1 GHz<br>$BW_{rel} \approx$ 40 % | slightly reduced $BW$ due to frequency dependent coupler presence and impedance level transformers |
| Pulsed-RF Measurements performed at $V_{DS}$= 25 V and 8.5 GHz | | | |
| $P_{out}$ | 39 dBm at $P_{in}$ = 31 dBm | 42 dBm at $P_{in}$ = 33 dBm | balanced approach delivers doubled output power with about the same PAE and the same power gain |
| PAE | 46 % | 44 % | |
| $G_P$ | 8 dB | 8 dB | |
| $G_{lin}$ | 11 dB | 10.2 dB | |

of the HPA Crianza becomes slightly reduced from 45 % to 40 % compared with the HPA Reserva. The reason for the reduced 3 dB bandwidth consists in the frequency dependent and limited bandwidth of the coupler structures and the additional impedance level transformation networks that are optimized only for a frequency range from 8 up to 10 GHz. In a final solution of the balanced AlGaN / GaN HEMT HPA design where the optimum load impedance level of the PA module ($\approx$ 25 Ω) would be directly matched to the 25 Ω output Lange coupler, the bandwidth limiting impedance transformers between the PA modules and the output coupler would be redundant. As a result, the bandwidth of the PA module should be preserved in the balanced approach. In addition, the final balanced design would require less MMIC chip size due to the eliminated impedance transformation networks.

Since two PA modules are used in parallel in the balanced design, almost twice the saturated output power of the single amplifier minus the net insertion losses of the couplers and impedance-transformation networks is achieved. This leads to a pulsed output power of almost 16 W at approximately 2 dB gain compression, 8.5 GHz, and at a drain bias voltage of $V_{DS}$ =25 V. The corresponding output power of the HPA Reserva amounts to about 8 W at a 3 dB gain compression level and at the same frequency and biasing conditions. These measurement results of the HPA Reserva and Crianza were obtained at an input power level of 31 dBm and 33 dBm, respectively. Thus, the balanced HPA needs almost twice the power supply to reach the doubled output power compared with the HPA Reserva. The associated power gain yields 8 dB for both designs. Finally, the obtained PAE of the HPA MMICs Reserva and Crianza at the above mentioned conditions are 46 % and 44 %, respectively. Accordingly, the HPA MMIC Crianza offers only a small decrease in power-added efficiency compared with the HPA Reserva.

Consequently, this demonstrator serves as a "proof-of-concept" for a first balanced microstrip AlGaN/GaN HEMT HPA MMIC using low-impedance Lange couplers. This modular concept delivers an excellent example of the advantages of using a balanced HPA design in GaN technology with highly improved small-signal matching capabilities and a doubled output power at the expense of only a few percent in the PAE performance. In a final optimized solution as described above, the performance of the balanced power amplifier should be further increased not only concerning the obtainable bandwidth and reduction of the required MMIC chip size, but also relating to the achievable output power, power gain, and efficiency of the circuit due to the reduced net insertion losses, especially on the output side of the high-power amplifier.

# Chapter 6

# High-Power Amplifiers Using Advanced Circuit Design Techniques

This chapter provides a refined overview of two different circuit design methods to optimize the PAE within AlGaN / GaN-HEMT power amplifier MMICs. In the first part, the impact of the driver (DRV)- to final (PA)-stage total gate-width (TGW) ratio on the PAE performance of a HPA is discussed in detail. The second part focuses on the effect of harmonic termination at the input and output of the HEMT-device on the PAE performance of the entire HPA.

## 6.1 Influence of PA/DRV TGW Ratio on PAE

Typical HPA specifications at X-band frequencies (8.5-11 GHz) include an output power target of $P_{out} \geq 20$ W at drain bias voltages of up to $V_{DS}$= 42 V. Such high $V_{DS}$-levels are not completely realizable at present because of the high power densities of the device with low PAE values at the same time leading to thermal problems concerning the structural-design technology (see Chapter 2.1). Therefore, the simulations and measurements of this chapter focus on drain bias voltages of $V_{DS}$= 30-35 V. Besides an output power beyond 20 W, a maximal possible PAE is desired for all HPA designs.

Hereby, a typical HPA gain specification of approximately 20 dB demands a dual-stage power-amplifier design consisting of a final (PA)-stage and a corresponding driver (DRV)-stage. Before considering the details of the circuit design, the specifications of the PA- and driver-stage size have to be determined on the basis of the performance of the used single power cell and the design goals of the entire HPA. Here, the purpose of the PA-stage mainly consists in the generation of the required output power of the entire HPA while the DRV-stage is responsible to provide the required driving power for the PA-stage transistors. Consequently, the most important aspect is to decide how much drive power is required in a particular case to "drive" the PA-stage into saturation [25].

Thus, this section deals with the influence of the PA/DRV TGW ratio on the PAE performance of dual-stage HPAs in theory and by means of practical HPA designs.

### 6.1.1 Selection of AlGaN/GaN Power Cell Geometry

To be able to obtain 20 W of output power with respect to a design reserve at the upper end of the frequency band of approximately 15 % for influencing factors such as variations between the HEMT model and the measured transistor and with respect to the net insertion loss of the output matching network ($\approx$ 1 dB), an output power of at least 44.6 dBm (28 W) is necessary. This output power is realizable with either four or eight power cells operated in parallel in the PA-stage. Since 8 power cells would yield to a very complex structure requiring a large chip size and leading to high losses within the dividing and combining networks according to [96], the HPA designs in this section have only four FET cells used in the PA-stage. Consequently, the single FET cell requires an output power of at least 7 W.

Therefore, the basic FET cell needs to be chosen and analyzed first. For simplification all DRV- and PA-stage transistors in this section refer to the same basic AlGaN/GaN HEMT cell. In this case, an 1 mm TGW basic FET cell is used with an $8\times125\,\mu$m gate geometry as a best compromise between gate-finger count and finger width (see Chapter 3, Section 3.1.1) to achieve optimum gain performance. The used HEMT device is based on the T-gate technology without source-connected shields and a gate-to-gate pitch of 50 $\mu$m. The aluminum concentration in the barrier is 22 %. From load-pull (LP) simulations of the HEMT as described in Chapter 3, the optimum load impedances of the transistor for maximum output power ($\Gamma_{L,Pout}$) and maximum PAE ($\Gamma_{L,PAE}$) were determined. The calculated load impedances strongly differ from each other as can be seen in Figure 6.1. The advantages of an obtainable maximum output power against an optimum PAE has to be weighed up before starting the actual HPA design.

The optimum load impedance for maximum output power yields $(29.6 + j24.2)\,\Omega$ at 10 GHz and 3 dB gain compression, while the optimum load impedance for maximum PAE results in $(13.5 + j38.6)\,\Omega$ at the same frequency. The associated optimum input impedance for maximum output power is $(3.4 + j7.7)\,\Omega$. Figure 6.2 depicts the simulated power-sweep of an AlGaN/GaN HEMT with $W_g$= 1 mm at 10 GHz terminated with $\Gamma_{L,Pout}$ (line without symbols) and with $\Gamma_{L,PAE}$ (line with symbols) and biased at $V_{DS}$= 30 V.

The device offers a linear gain of 13.5 dB in the case of a termination with $\Gamma_{L,Pout}$. The simulated output power at 3 dB gain compression (marked in the figure) yields about 38.6 dBm, i.e. 7.3 W/mm, and the associated PAE reaches 54.1 %. In the case of a termination with $\Gamma_{L,PAE}$, an 1.5 dB higher linear gain of approximately 15 dB is reached. The comparable values for $P_{out}$, PAE and power gain at the same input power level as before yield 36.4 dBm, i.e. 4.5 W/mm, 64 %,

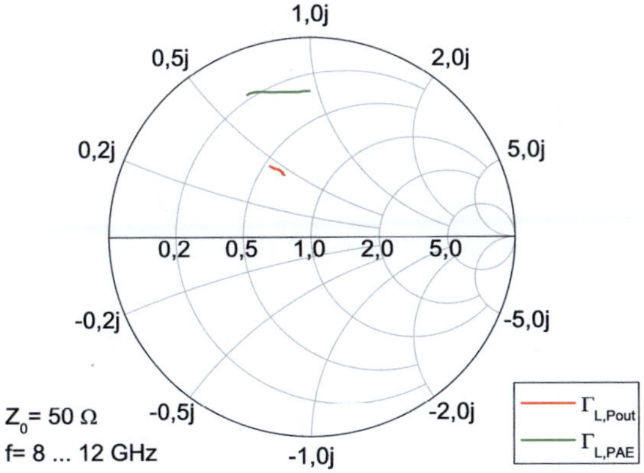


Figure 6.1: Load-Pull simulated optimum load impedances of a 1 mm AlGaN/GaN HEMT for maximum output power and PAE from 8-12 GHz. Bias point: $V_{DS}$= 30 V and $V_{GS}$= -2.25 V.

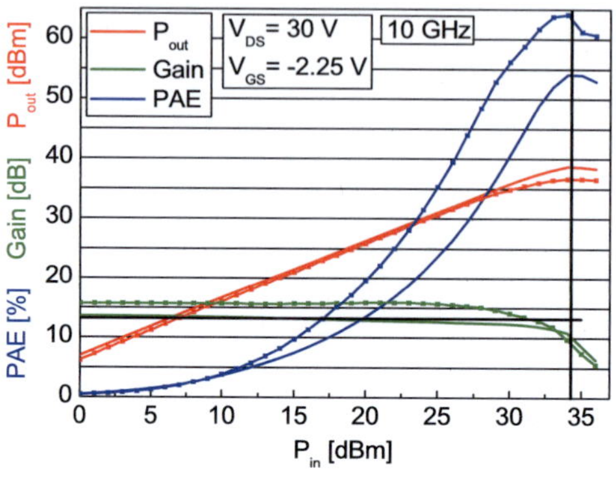


Figure 6.2: CW-simulated power-sweep of a 1 mm AlGaN/GaN HEMT at 10 GHz terminated with the optimum $\Gamma_{L,Pout}$ (line without symbols) and $\Gamma_{L,PAE}$ (line with symbols). Bias point: $V_{DS}$=30 V and $V_{GS}$=-2.25 V.

and 10.8 dB, respectively. Thus, the PAE can be improved by 10 % on device level at the expense of a strongly reduced output power.

Accordingly, the required 20 W of output power can just be realized with four of the above mentioned power cells combined in the PA-stage and terminated with $\Gamma_{L,Pout}$ at the expense of less power-added efficiency. This result fits to the experimental investigation of this $8 \times 125\,\mu$m basic FET cell via LP measurements and confirms the above theoretical considerations about the selection of the optimum cell geometry for the HPA designs.

### 6.1.2 Analytical Calculations of PA/DRV TGW Dependency

For a fundamental analytical investigation, the calculation of the PAE of a dual-stage HPA is divided into three subsequent approaches of increasing complexity:

1. Strongly Simplified Analysis

2. Refined Example with "real" Compression Levels

3. Realistic Example.

The assumptions made in the three analysis steps are summarized in Figure 6.3. In the first approach the influencing factors of the PAE are assessed by using strongly simplified assumptions for the HPA system to achieve PAE calculations via analytic formulae. Within the next two steps, these strongly simplified assumptions are replaced by more realistic descriptions to finally describe a real dual-stage HPA case. With the help of this step-by-step analysis of the PAE, the main influencing factors of the PAE are presumed to be identified and theoretically evaluated.

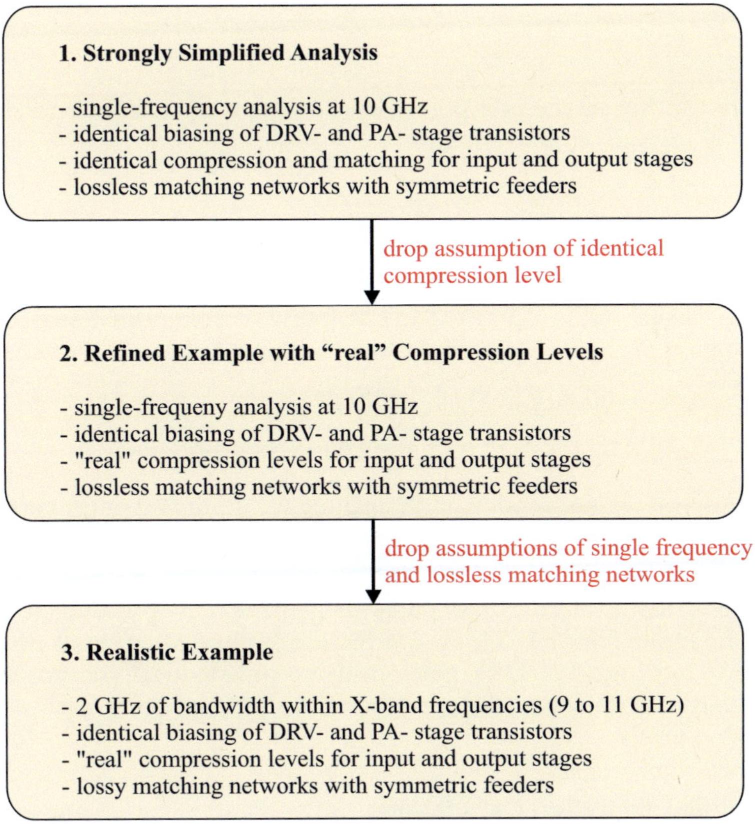


Figure 6.3: Assumptions of the three analysis-steps for a sophisticated PAE calculation.

Figure 6.4 shows the schematic of a of a power-level diagram of a dual-stage HPA with its dependencies of DRV- and PA-stage as well as all other characteristics of the amplifier mentioned in the following equations. Feeders to the individual FET cells are included to the passive networks, in particular into the IMN, ISMN, and OMN.

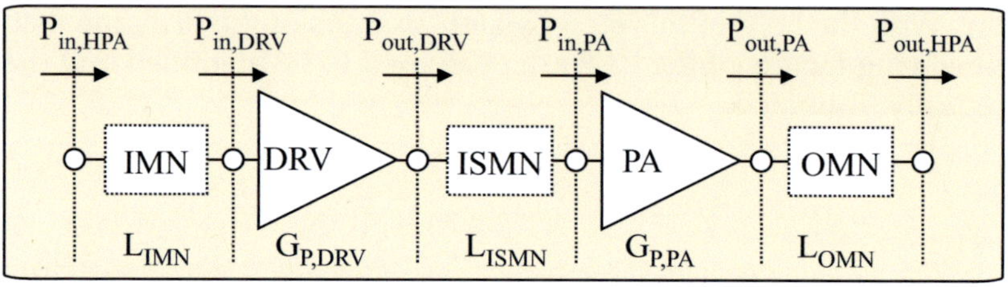


Figure 6.4: Schematic of a power-level diagram of a dual-stage HPA with all relevant characteristics.

## 1. Strongly Simplified Analysis

In a first analysis, some assumptions are made to assess the PAE influencing factors with respect to the possibility of a simplified analytical calculation of the PAE according to a varying PA/DRV TGW ratio:

- single-frequency analysis at 10 GHz
- identical biasing of DRV- and PA- stage transistors
- identical compression and matching for input and output stages
- lossless matching networks with symmetric feeders.

Here, identical compression is in general only possible at one specific input power level, or approximately for a very small power sweep. For high PA/DRV gate-width ratios, identical compression normally occurs close to the saturated output power at high compression levels.

### Simplified Analytical Calculations

Neglecting any network losses and assuming the same loadline for the driver- and PA-stage, the output power of both HPA stages only depends on the number of FET cells used in each stage ($N_{\mathrm{FET/DRV}}$

and $N_{FET/PA}$) and the output power of the single FET cell provided at the above-mentioned assumptions. Thus, the output power of both HPA stages can be calculated according to Equations 6.1 and 6.2.

$$P_{out,PA} = N_{FET/PA} \cdot P_{out,FET} \tag{6.1}$$

$$P_{out,DRV} = N_{FET/DRV} \cdot P_{out,FET} \tag{6.2}$$

Since the compression behavior of both HPA stages is not taken into account in this strongly simplified analysis, no lower limit exists for the DRV-stage size based on the above-stated assumptions. But it has to be considered that in a real case, the output power of the driver-stage minus the losses of the interstage matching network indicates the lowest value that is required to "feed" the PA-stage. This output power is likewise necessary to prevent the DRV-stage transistors from going into compression before the PA-stage transistors.

However, the relationship between the possible output power of the DRV- and PA-stage of a HPA directly depends on the ratio of the total gate-width (TGW) of both stages leading to equation

$$\frac{P_{out,DRV}}{P_{out,PA}} = \frac{TGW_{DRV}}{TGW_{PA}} \; . \tag{6.3}$$

Assuming all matching networks being lossless, the ratio of the DRV- and PA-stage output power can also be shown as

$$\frac{P_{out,DRV}}{P_{out,PA}} = \frac{P_{in,PA}}{P_{out,PA}} = \frac{1}{G_{P,PA}} \tag{6.4}$$

leading to the following equation:

$$G_{P,PA} = \frac{TGW_{PA}}{TGW_{DRV}} \; . \tag{6.5}$$

The latter equation holds only for a small input power segment of the power sweep close to PAE-peak. For $P_{out,PA} / P_{in,PA} <$ "$P_{opt}$", $G_{P,PA}$ is larger than $TGW_{PA} / TGW_{DRV}$, while $G_{P,PA}$ turns out to be smaller than $TGW_{PA} / TGW_{DRV}$ for $P_{out,PA} / P_{in,PA} >$ "$P_{opt}$".

On closer examination of the power-added efficiency and its dependence on both amplifier stages, the PAE is calculated in detail for the entire HPA as well as for both transistor stages. In this first step, all matching networks are assumed to be lossless for simplification.

The particular PAEs of both power amplifier stages are calculated as follows referring to the basic PAE-equation from Chapter 3 (Equation 3.4) and the above-mentioned equations and simplified assumptions:

$$\begin{aligned} PAE_{\mathrm{DRV}} &= \frac{P_{\mathrm{out,DRV}} - P_{\mathrm{in,DRV}}}{P_{\mathrm{DC,DRV}}} \\ &= \frac{P_{\mathrm{out,DRV}}}{N_{\mathrm{FET/DRV}} \cdot P_{\mathrm{DC,FET}}} \cdot \left(1 - \frac{1}{G_{\mathrm{P,DRV}}}\right) \end{aligned} \tag{6.6}$$

$$\begin{aligned} PAE_{\mathrm{PA}} &= \frac{P_{\mathrm{out,PA}} - P_{\mathrm{in,PA}}}{P_{\mathrm{DC,PA}}} \\ &= \frac{P_{\mathrm{out,PA}}}{N_{\mathrm{FET/PA}} \cdot P_{\mathrm{DC,FET}}} \cdot \left(1 - \frac{1}{G_{\mathrm{P,PA}}}\right) . \end{aligned} \tag{6.7}$$

For the entire HPA, the PAE yields

$$PAE_{\mathrm{HPA}} = \frac{P_{\mathrm{out,HPA}}}{P_{\mathrm{DC,DRV}} + P_{\mathrm{DC,PA}}} \cdot \left(1 - \frac{1}{G_{\mathrm{P,HPA}}}\right) . \tag{6.8}$$

Provided that the DC power is independent of the size of each transistor stage based on the assumption of identical biasing of the DRV- and PA- stage transistors, the equation above (6.8) can be restated depending on $P_{\mathrm{DC,FET}}$. As a result of the assumed DC power simplification, $P_{\mathrm{DC}} = V_{\mathrm{DC}} \cdot I_{\mathrm{DC}}$ is assumed to be constant as long as $I_{\mathrm{DC}}$ is constant which is the case for Class-A operation. In addition, the assumption of identical compression behavior leads to $G_{\mathrm{P,DRV}} = G_{\mathrm{P,PA}} = G_{\mathrm{P,FET}}$.

$$
\begin{aligned}
PAE_{\text{HPA}} &= \frac{P_{\text{out,HPA}}}{(N_{\text{FET/DRV}} + N_{\text{FET/PA}})P_{\text{DC,FET}}} \cdot (1 - \frac{1}{G_{\text{P,HPA}}}) \\
&= \frac{P_{\text{out,HPA}}}{(1 + \frac{TGW_{\text{DRV}}}{TGW_{\text{PA}}})P_{\text{DC,PA}}} \cdot (1 - \frac{1}{G_{\text{P,DRV}}G_{\text{P,PA}}}) \\
&= DE_{\text{FET}} \cdot \frac{1}{(1 + \frac{TGW_{\text{DRV}}}{TGW_{\text{PA}}})} \cdot (1 - \frac{1}{{G_{\text{P,FET}}}^2}) . \qquad (6.9)
\end{aligned}
$$

Equation 6.9 shows that as the $TGW_{\text{DRV}}$ decreases the PAE of the entire amplifier increases. As a result, the DRV periphery needs to be designed as small as possible concerning to an optimized $PAE_{\text{HPA}}$. Nevertheless, it still has to be large enough to "drive" the PA-stage.

**Influence of PA/DRV TGW ratio on PAE in Simplified Example**

In Equation 6.9 the PAE performance of a high-power amplifier is shown only in dependency on the PA/DRV TGW ratio of the HPA itself and the performance of the unit FET cell that is described in Section 6.1.1. Since the number of FETs used in the PA-stage is determined as 4 being necessary to produce the required typical output power of 20 W of the HPA, only the number of FET cells used in the DRV-stage is variable. According to Figure 6.2, the output power of the entire HPA is equal to the output power of the PA-stage in case of a lossless output matching network, i.e. the saturated $P_{\text{out,HPA}} = 4 \cdot P_{\text{out,FET}} = 44.6\,dBm$.

With the above-mentioned assumptions, the power gain at 3 dB compression level equals 10.5 dB for the single FET cell (see Figure 6.2) as well as for N parallel FET cells leading to a power gain of the PA- and driver-stage at 3 dB compression of 10.5 dB as well. Consequently, the power gain of the entire dual-stage HPA amounts to $G_{\text{P,HPA}} = {G_{\text{P,FET}}}^2 \mathrel{\widehat{=}} 21$ dB. According to Figure 6.2, the drain efficiency of the FET cell is $DE_{\text{FET}} = 59.6\,\%$.

Figure 6.5 shows the PAE dependency on the PA/DRV GW ratio by including these values in Equation 6.9 for different driver-stage sizes and a DC power that is only dependent on $P_{\text{in}}$, but independent of $TGW_{\text{DRV}}$. The figure shows the simplified calculated PAE-sweeps of HPAs with different PA / DRV TGW ratios at 10 GHz and $V_{DS}$= 30 V as well as a PAE-sweep of the CW-simulated 1 mm AlGaN / GaN power cell terminated with the optimum $\Gamma_{L,Pout}$ from Figure 6.2 at the same frequency and biasing for comparison.

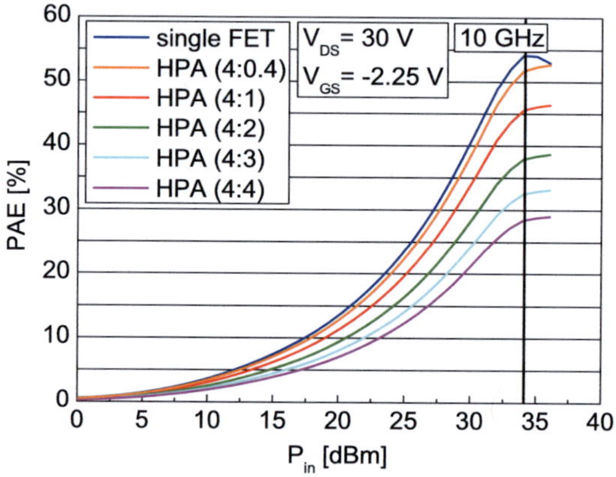


Figure 6.5: CW-simulated PAE-sweep of a 1 mm AlGaN / GaN HEMT as well as simplified calculated PAE-sweeps of HPAs according to Equation 6.9 with different PA / DRV TGW ratios at 10 GHz. Bias point: $V_{DS}$= 30 V and $V_{GS}$= -2.25 V.

The theoretical upper limit of a realizable PAE of a power amplifier ($PAE_{HPA}$) amounts to 51.7 % for a PA / DRV TGW ratio of 4:0.4. The ratio 10:1 is chosen on the basis of the theoretical PA driving limit that nearly exactly fits to $G_{P,HPA}$= 10.5 dB per stage (or FET). In a real HPA design, a much lower PA / DRV TGW ratio is required due to the network losses and bandwidth limitations, especially caused by the ISMN mismatches. For a PA / DRV TGW ratio of one the theoretically lowest $PAE_{HPA}$ value of 28.4 % is obtained, demonstrating a reduction by nearly 50 % compared with the PAE of the single FET cell.

**Conclusion of Simplified Assumptions**

As expected from Equation 6.9, the PAE of the entire amplifier increases with increasing PA / DRV TGW ratio, that means with decreasing driver-stage size at a constant PA-stage size. Table 6.1 summarizes the calculated PAE values for the HPAs from Figure 6.5 at 3 dB gain

Table 6.1: CW-simulated PAE of a $W_g$=1 mm FET cell as well as simplified calculated PAE of HPAs with different PA / DRV TGW ratios at 3 dB gain compression, $V_{DS}$=30 V and 10 GHz according to Equation 6.9 and Figure 6.5.

| PA/DRV TGW Ratio | Gain/transistor-stage | $P_{out,sat}$ | PAE |
|---|---|---|---|
| single FET cell | 10.5 dB | 38.6 dBm | 54.1 % |
| 4:0.4 (10:1) | 10.5 dB | 44.6 dBm | 51.7 % |
| 4:1 | 10.5 dB | 44.6 dBm | 45.5 % |
| 4:2 | 10.5 dB | 44.6 dBm | 37.9 % |
| 4:3 | 10.5 dB | 44.6 dBm | 32.5 % |
| 4:4 (1:1) | 10.5 dB | 44.6 dBm | 28.4 % |

compression ($P_{in}$=34 dBm) performed at 10 GHz with a Class-AB biasing of $V_{DS}$=30 V and $V_{GS}$=-2.25 V. For comparison, the performance of the single FET cell according to Figure 6.2 is recorded in the table as well.

As can be seen in the figure and table, with decreasing driver-stage size, the theoretical upper limit of the possible $PAE_{HPA}$ is reached with a PA / DRV TGW ratio of 10:1 and amounts to 51.7 %. Furthermore, a PA / DRV TGW ratio of one leads to a reduction of the maximum PAE value by nearly 50 % compared to the single FET cell. In reality, this is typical for a Class-A operation where the required DC power is constant. Nevertheless, in this case the interaction of all the simplified assumptions leads to this strongly reduced PAE value.

Consequently, the maximum possible PAE of a dual-stage power amplifier will be in the range of the approximately 50 % up to 95 % of the maximum PAE value of the unit transistor cell depending on the PA / DRV TGW ratio and under consideration of the simplified assumptions to the HPA defined above without any network losses. According to the above stated results, it seems that the PA / DRV TGW ratio has a very strong effect on the PAE performance of a dual-stage HPA that will be relativized within the next analysis steps with more realistic assumptions according to a real HPA design.

**2. Refined Example with "Real" Compression Levels**

In the next analysis step, the "identical" compression levels for the input and output stages are replaced by "real" compression levels depending on the size of each HPA stage. All other assumptions are the same as in the first analysis step:

- single-frequency analysis at 10 GHz
- identical biasing of DRV- and PA- stage transistors
- **"real" compression levels for input and output stages**
- lossless matching networks with symmetric feeders.

The compression behavior of both amplifier stages is a very important factor concerning the realizable PAE. The conclusion of the simplified analysis is that the PAE of a dual-stage HPA can be optimized by minimizing the driver stage size. This is only true for the assumption of an identical compression behavior of both stages. In real cases, the compression levels of the input and output stage strongly depend on the size of each transistor stage. On scaling down the driver stage size, the transistors of the input stage are driven more and more into compression while still being able to "drive" the PA-stage transistors. In this case, the calculated output power of the DRV-stage indicates the lowest value that is required to "feed" the PA-stage assuming lossless matching networks leading to Equation 6.10:

$$P_{\mathrm{out,DRV}} \geq \frac{P_{\mathrm{out,PA}}}{G_{\mathrm{P,PA}}} \,. \tag{6.10}$$

This output power is likewise necessary to prevent the DRV-stage transistors from going into compression before the PA-stage transistors. However, a large driver does not go into compression before the PA-stage transistors but wastes a lot of DC power that is not converted into driving power for the PA-stage leading to a reduced PAE of the entire HPA.

Figure 6.6 displays the CW-simulated output power of a driver-stage with one or two parallel basic FET cells as well as the CW-simulated input power of a PA-stage with four parallel FET cells. All power-sweeps are performed at 10 GHz with a Class-AB biasing of $V_{\mathrm{DS}}$=30 V and $V_{\mathrm{GS}}$=-2.25 V.

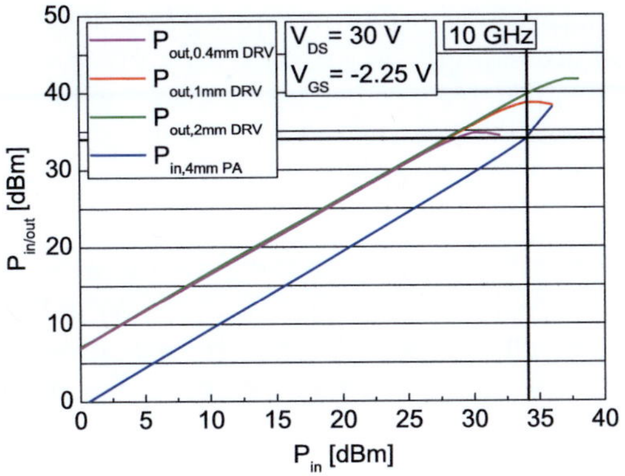


Figure 6.6: CW-simulated output power of a DRV-stage with one and two basic FET cells and CW-simulated input power of a PA-stage with 4 parallel FET cells, each at 10 GHz and at $V_{DS}$= 30 V and $V_{GS}$= -2.25 V.

According to Equation 6.10, the output power of the driver has to be larger than the required input power of the power-stage to be able to drive the PA-stage. For reaching a saturated output power of the HPA, the 4 mm PA stage needs an input power of 34.1 dBm ($P_{out,PA/sat} / G_{P,PA}$) as marked in the figure.

With regard to Figure 6.6, even a driver consisting of only 0.4 mm TGW (PA/DRV TGW ratio of 10:1) is just able to drive a 4 mm power-stage up to the optimum PAE point of the power sweep. In this case, the special driver condition $G_{P,PA} = G_{P,DRV} = TGW_{PA} / TGW_{DRV}$ is valid only in the point of simultaneous saturation of the PA and DRV stage with a minimal gain reserve of 0.5 dB or DRV stage power, respectively. However, this example does not include any network losses and is assumed to operate only at the single-frequency of 10 GHz. Additional network losses and at least a narrow bandwidth around 10 GHz lead to an increase of the required driver-stage output power.

According to the figure, the 1 mm DRV-stage delivers 4.5 dBm more of saturated output power than the required minimum input-power level of 34.1 dBm for the PA stage. Therefore, a DRV-stage with at least 1 mm gate-periphery delivers a more practical "output power" value that is necessary for a real HPA design with a 4 mm PA stage with network losses.

Taking the required output- and input-power level of the PA-stage of 44.6 dBm and accordingly 34.1 dBm with a resulting PA-stage gain of 10.5 dB into account, a more precise calculation based on Equation 6.8 can be done under consideration of a "real" compression level and a realistic DC-power consumption of the DRV-stage depending on its total gate-width.

Table 6.2 gives an overview of the PAE behavior in the refined example for HPAs with different PA/DRV TGW ratios at 3 dB gain compression, $V_{DS}$=30 V and 10 GHz. In each case, the values for power gain and associated drain efficiency of the driver at the particular input-power level are taken from Figure 6.2 and declared in Table 6.2.

The following conclusions can be drawn out of the results shown in Table 6.2. On the one hand, the larger the DRV-stage size, the less the transistors of the input stage are driven into compression leading to a higher power gain of the driver-stage and an insignificant increase in PAE of the entire HPA according to Equation 6.8. On the other hand, the larger the driver-stage size, the higher is the DC-power consumption of the driver and consequently of the entire HPA resulting in a strong decrease in PAE. The influence of a real compression behavior and a realistic DC-power consumption of the driver-stage becomes important, especially with decreasing PA/DRV TGW ratio when comparing Table 6.1 with Table 6.2. In the latter example, the achievable PAE values of the HPAs with different PA/DRV TGW ratios are between 81 % and 96 % of the maximum PAE value

Table 6.2: Refined calculated PAE of HPAs with different PA/DRV TGW ratios at 3 dB gain compression, $V_{DS}$=30 V and 10 GHz according to Equation 6.8 under consideration of a "real" compression level and DC-power consumption of the DRV-stage depending on its TGW.

| **PA/DRV TGW Ratio** | $DE_{\mathbf{ass,DRV}}$ | $G_{\mathbf{P,DRV}}$ | $P_{\mathbf{DC,DRV}}$ | $P_{\mathbf{DC,HPA}}$ | $PAE_{\mathbf{HPA}}$ |
|---|---|---|---|---|---|
| 4:0.4 (10:1) | 57 % | 11.5 dB | 4.55 W | 53.55 W | 52 % |
| 4:1 | 36.2 % | 12.4 dB | 7.16 W | 56.16 W | 49.7 % |
| 4:2 | 25 % | 12.5 dB | 10.38 W | 59.38 W | 47 % |
| 4:3 | 20.7 % | 12.65 dB | 12.49 W | 61.49 W | 45.4 % |
| 4:4 (1:1) | 18.3 % | 12.7 dB | 14.16 W | 63.16 W | 44.2 % |

of the unit transistor cell. Again, the networks are assumed to be lossless.

**3. Realistic Example with Real Bandwidth and Lossy Matching Networks**

In a final analysis step, the HPA is presumed to be operated over a 2 GHz bandwidth within X-band frequencies from 9 to 11 GHz. In addition, the frequency-dependent losses of the matching networks are taken into account. The assumptions made for this example are specified as follows:

- **2 GHz of bandwidth within X-band frequencies (9 to 11 GHz)**
- identical biasing of DRV- and PA- stage transistors
- "real" compression levels for input and output stages
- **lossy matching networks with symmetric feeders**.

For a frequency-dependent gain analysis, load-pull simulations of the single FET cell are done over the entire X-band frequency range from 8 up to 12 GHz. Figure 6.7 depicts CW-simulated power gain-sweeps of an 8x125 $\mu$m AlGaN/GaN power cell for different input power levels over X-band frequencies with a drain and gate biasing of $V_{DS}$=30 V and $V_{GS}$=-2.25 V, respectively.

The "small-signal" gain curve at $P_{in}$=0 dBm shows a strong decrease in gain with increasing frequency from approximately 16 dB at the lower end of the X-band down to 11.5 dB at the upper end of the X-band. The frequency-dependent gain slope characteristics stay constant up to an input power level of 30 dBm and then change drastically over the frequency range for even higher large-signal input power levels.

The relatively high losses towards the upper end of the required HPA frequency bandwidth define the upper limit for the gate-width for the basic FET cells within the output stage. According to the above-stated assumptions, the HPA is designed to be operated with a flat gain for a frequency bandwidth from 9 to 11 GHz. To compensate the strongly reduced power gain at the upper end of the frequency band with the objective to obtain gain flatness over the desired frequency range, the input and interstage matching networks are designed and optimized for best matching at the upper end of the

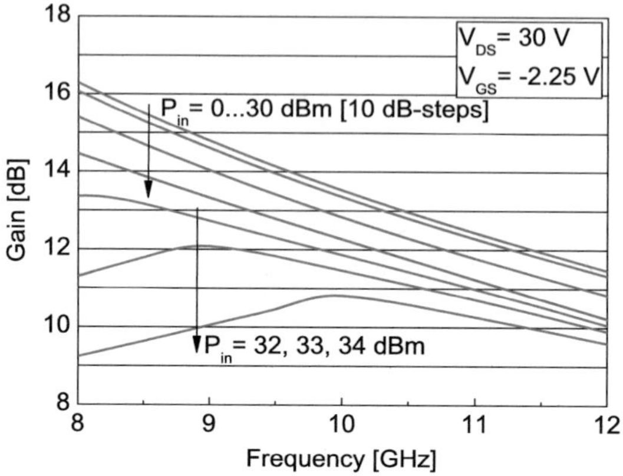


Figure 6.7: CW-simulated power gain-sweeps of a 1 mm AlGaN/GaN HEMT for different input power levels at X-band frequencies. Bias point: $V_{DS}$=30 V and $V_{GS}$=-2.25 V.

frequency band. In addition, losses in the ISMN for lower frequencies may be introduced intentionally.

Consequently, the low power gain at the upper end of a specific frequency bandwidth as well as the gain compression at the operating input power level and the losses of the passive matching networks strongly affect the input power of the PA-stage. For a 2 GHz bandwidth from 9 up to 11 GHz, the power gain at 11 GHz at an appropriate power level has to be taken into account. With regard to Figure 6.7, this power gain yields approximately 11 dB. Depending on the frequency and the number of FET cells used in parallel in each transistor stage, typical net insertion losses for the input and interstage matching networks are approximately 1.5-2 dB per network. In a worst-case scenario, the power gain of the PA-stage at the upper end of the frequency band would result to only 7 dB.

Before delving into the real HPA design, the influencing factors on both HPA-stages are considered and some realistic analytical calculations are done in terms of the PAE.

**Key Issues to the PA/DRV TGW Ratio**

The determination of the PA-stage periphery primarily depends on the following three factors:

- the required output power of the entire power amplifier $P_{\mathrm{out,HPA}}$
- the power density of the basic FET cell $P_{\mathrm{out,FET/PA}}$ used in the PA-stage
- the loss of the output matching network $L_{\mathrm{OMN}}$.

Besides these obvious factors, the output power variation of the FET cells resulting from the process and heatsink variations should be considered and incorporated into an additional output power margin $C_{\mathrm{P,add/PA}}$ of the PA-stage. Consequently, the number of transistors $N_{\mathrm{FET/PA}}$ used in parallel in the PA-stage can be calculated as follows [29]:

$$N_{\mathrm{FET/PA}} \geq \frac{P_{\mathrm{out,HPA}}}{P_{\mathrm{out,FET/PA}} \cdot L_{\mathrm{OMN}} \cdot C_{\mathrm{P,add/PA}}}. \tag{6.11}$$

In this case, the output power that is available for the unit transistor cell depends on the required bias operation and frequency-range as well as on the transistor topology.

Similar considerations as for the PA-stage can be done for the driver. In this case, the size of the driver-stage transistors mainly depends on

- the required total output power of the first stage $P_{\mathrm{out,DRV}}$
- the number of transistors used in parallel in the DRV-stage
- the power gain of the PA-stage transistors $G_{\mathrm{P,FET/PA}}$ and
- the expected loss of the interstage matching network $L_{\mathrm{ISMN}}$.

Together with the above mentioned factors, an extra output power margin $C_{\mathrm{P,add/DRV}}$ has to be provided for output power variations due to process and heatsink variations as well as to prevent the DRV-stage transistors from going into compression before the PA-stage transistors. Taking all these factors into account, the required output power of the DRV-stage transistors results in equation

$$P_{\mathrm{out,DRV}} = \frac{N_{\mathrm{FET/PA}} \cdot P_{\mathrm{out,FET/PA}}}{G_{\mathrm{P,PA}} \cdot L_{\mathrm{ISMN}}} \cdot C_{\mathrm{P,add/DRV}} \,. \tag{6.12}$$

**Realistic Analytical Calculations**
The implicit total power-added efficiency of the HPA is calculated with the help of the simplified PAE-equation 6.8 but in consideration of the losses of the matching networks. Thus, the $PAE_{\mathrm{HPA}}$ yields to

$$\begin{aligned} PAE_{\mathrm{HPA}} &= \frac{P_{\mathrm{out,HPA}}}{P_{\mathrm{DC,DRV}}+P_{\mathrm{DC,PA}}} \cdot \left[1-\frac{1}{G_{\mathrm{P,DRV}}G_{\mathrm{P,PA}}L_{\mathrm{IMN}}L_{\mathrm{ISMN}}L_{\mathrm{OMN}}}\right] \\ &= \frac{P_{\mathrm{in,HPA}}}{P_{\mathrm{DC,DRV}}+P_{\mathrm{DC,PA}}} \cdot \\ &\quad \cdot \left[G_{\mathrm{P,DRV}}G_{\mathrm{P,PA}}L_{\mathrm{IMN}}L_{\mathrm{ISMN}}L_{\mathrm{OMN}}-1\right]. \end{aligned} \tag{6.13}$$

When $P_{\mathrm{DC,DRV}}$ and $P_{\mathrm{DC,PA}}$ of Equation 6.13 are replaced with the solutions from Equations 6.6 and 6.7 and with respect to the losses of the matching networks, the power-added efficiency of the entire HPA results in

$$PAE_{\mathrm{HPA}} = \frac{PAE_{\mathrm{DRV}}PAE_{\mathrm{PA}}\left[L_{\mathrm{IMN}}L_{\mathrm{ISMN}}L_{\mathrm{OMN}}G_{\mathrm{P,DRV}}G_{\mathrm{P,PA}}-1\right]}{PAE_{\mathrm{DRV}}L_{\mathrm{IMN}}L_{\mathrm{ISMN}}G_{\mathrm{P,DRV}}\left[G_{\mathrm{P,PA}}-1\right]+\cdots} \\ \overline{\cdots+PAE_{\mathrm{PA}}L_{\mathrm{IMN}}\left[G_{\mathrm{P,DRV}}-1\right]}\,. \tag{6.14}$$

There exists no explicit solution for this equation. According to [29] the assumption of $G_{\mathrm{P,DRV}}$ and $G_{\mathrm{P,PA}} >> 1$, reduces Equation 6.14 to

$$PAE_{\mathrm{HPA}} \approx PAE_{\mathrm{PA}} \cdot L_{\mathrm{OMN}}\,. \tag{6.15}$$

The assumption of comparatively high power gains of the DRV- and PA-stage compared to one is just a special case that cannot be fulfilled over the required 2 GHz frequency range from 9 up to 11 GHz. This is because of the gain compression losses at the operating input power levels together with the high losses of the power gain towards the upper end of the frequency range (see Figure 6.7) and the net insertion losses of the passive matching networks. Nevertheless, the power-added efficiency of the entire HPA primarily depends on the PAE of the power-stage and the net insertion losses within the output matching network. This conclusion is in agreement with the main purpose of the DRV-stage to generate maximum gain at appropriate power levels over the entire bandwidth while having a relatively low impact of driver efficiency in accordance with [25].

**Conclusion of the Analytical Calculations**
Besides the PAE influencing factors out of Equation 6.15, the PAE of the entire HPA is also strongly affected by the PA/DRV gate-width ratio according to Equation 6.9. Up to now it was assumed that only one unit FET cell size is used within the HPA design for the driver- and PA-stage, i.e. a 1 mm basic FET cell with an $8\times125\,\mu$m gate geometry. In the first and simplest theory presented in this section, only the total cell size has an effect on the RF-PAE behavior of the DRV and PA stages. Together with practical gain versus frequency considerations (MSG/MAG characteristics), the cell geometry comes into play, see above Subsection 6.1.1. In theory, a driver with a gate-width of 2 mm can be designed with two of the above mentioned 1 mm basic FET cells used in parallel, or with a single larger FET cell with a gate-width of 2 mm. In this case, the MAG degradation of the FET cell with the enlarged gate-width has to be analyzed and compared to the lower net insertion losses of the passive matching networks concerning the reduction of the combining network that is necessary for the two parallel operated FET cells. This trade-off between the MAG degradation of the single FET cell depending on the gate-width of the transistor and the impact on the passive losses when combining various FET cells is a further design aspect that should not be ignored. In practice, it turns out that the MSG/MAG characteristic of the 2 mm FET cell is unacceptable for X-band frequencies. The practical selection should be $2\times1$ mm cells in this case.

Concluding, the PA-stage has to be designed for operation at best efficiency or maximum output power depending on the HPA specifications, while the driver-stage needs to be designed with the lowest periphery that is required to "feed" the PA-stage and simultaneously to prevent the DRV-stage transistors from going into compression before the PA-stage transistors. Enlarging the TGW of the driver leads to a lower compression level and therefore to an increased power gain at the expense of more dissipated DC power that is not translated into output power by the HPA FETs. This leads to a reduced overall efficiency.

Based on the above given analytical calculations, the optimization of the HPA driver-stage size has a significant impact on the entire HPA design. On the one hand, it strongly affects the overall efficiency of the HPA. On the other hand, the DRV-stage gate-width mainly determines the realizable bandwidth of the power amplifier.

Table 6.3 summarizes the theoretical possible PAE values of high-power amplifiers with different PA/DRV TGW ratios, each at the same compression level, frequency, and with the same biasing for good comparison. In addition, the PAE of a 1 mm AlGaN/GaN power cell at the same operation is recorded in the table.

As already mentioned before, the PAE values according to Table 6.1 give a first impression of the influence of the PA/DRV gate-width ratio on the PAE performance of a dual-stage HPA. Nevertheless, the values strongly differ for varying PA/DRV TGW ratios in this simplest calculation leading to unrealistic PAE results.

In the second case, the same Equation (Eq. 6.8) is taken for the PAE calculation, but with modified assumptions, i.e. a more realistic compression level and DC-power consumption of the DRV-stage depending on its total gate-width. As well as for the first and simplest calculation, the passive matching networks are assumed to be lossless. Here, the PAE results still depend on the PA/DRV TGW ratio, but seem to be more realistic being in the range of approximately 44 % up to 52 % based on a single FET cell with a PAE of 54.1 %.

The third PAE-column contains the calculated PAE according to Equation 6.13 that differs from Equation 6.8 in the form that it takes the losses of the three passive matching networks into account. These losses are estimated to -2 dB for the IMN as well as for the ISMN and -0.5 dB for the OMN at the single frequency of 10 GHz. Therefore, the last column delivers the most realistic PAE-values for the different

Table 6.3: CW-simulated PAE of a 1 mm AlGaN/GaN power cell as well as analytical calculated PAE of HPAs with different PA/DRV TGW ratios at 3 dB gain compression ($P_{in}$= 34 dBm), at 10 GHz, biased in Class-AB operation, and with the assumptions of the three analysis steps defined above.

| **PA/DRV TGW Ratio** | $P_{\mathbf{out,FET/HPA}}$ | $PAE_{\mathbf{FET/HPA}}$ **(Tab. 6.1)** | $PAE_{\mathbf{FET/HPA}}$ **(Tab. 6.2)** | $PAE_{\mathbf{FET/HPA}}$ **(Eq. 6.13)** |
|---|---|---|---|---|
| single FET | 38.6 dBm | 54.1 % | 54.1 % | 54.1 % |
| 4:0.4 (10:1) | 44.6 dBm | 51.7 % | 52 % | 51.4 % |
| **4:1** | 44.6 dBm | 45.5 % | 49.7 % | **49.1** % |
| **4:2** | 44.6 dBm | 37.9 % | 47 % | **46.5** % |
| 4:3 | 44.6 dBm | 32.5 % | 45.4 % | 44.9 % |
| 4:4 (1:1) | 44.6 dBm | 28.4 % | 44.2 % | 43.7 % |

PA/DRV TGW ratios at the single frequency of 10 GHz. The comparison of the last two columns shows that the net insertion losses of the passive matching networks diminish the PAE performance of the HPA at about 0.5-0.6 %. The two PA/DRV TGW ratios that are considered in the following are bolded in the table.

### 6.1.3 HPA Designs with Focus on PA/DRV TGW Ratio

In the following, the importance of DRV-stage size will be discussed with respect to PAE improvements by means of two dual-stage HPA designs with different DRV-to-PA-stage gate-width ratios. The design goals for both amplifiers are the same, namely a saturated output power of $P_{sat} \approx 20$ W with a PAE target of 40 % under approximately 3 dB gain compression and up to 2 GHz of bandwidth within X-band frequencies, i.e. 20 % of the relative bandwidth $BW_{rel}$.

An output power of 20 W can be achieved with the use of four 8x125 $\mu$m AlGaN/GaN power cells combined in the PA-stage yielding a PA-stage periphery of 4 mm as described in detail in Subsection 6.1.1. Thus, the determined PA-stage periphery is used for both HPA MMICs.

**Power Level Diagram**

Figure 6.8 gives an overview of a dual-stage HPA with different PA/DRV gate-width ratios with respect to the required output power of the entire MMIC. The figure should be examined from the output to the input. First of all, the design goal of a saturated output power of $P_{sat} \approx 20$ W needs to be placed at the output of the amplifier. With a simulated net insertion loss of nearly 0.5 dB for the output matching network including the 4:1 power combiner as well as with a PA-stage gain of 10.5 dB, the required PA-stage input power needs to be at least 33 dBm. Furthermore, considering a typical net insertion loss of approximately 2 dB for an interstage matching network which primarily depends on the required frequency range and the number of combiners and dividers of the unit transistors used in the DRV- and PA-stage, the required driver output power is approx. 8.5 dB lower than the PA output power, i.e. 35 dBm. With some reserve for broadband performance, the output power of the DRV-stage needs to be large enough ($\geq 35$ dBm) to enable a saturated output power of $P_{sat} \approx 20$ W over the full bandwidth of 9-11 GHz.

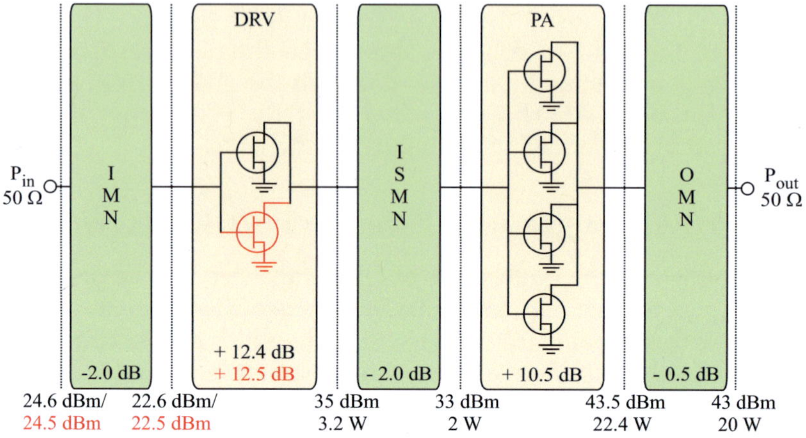


Figure 6.8: Block diagram of a dual-stage amplifier with different output/driver gate-width ratios.

### Determination of the DRV Periphery

Depending on the results from the analytical calculations, the size of the driver-stage has to be determined. For simplification, within the driver-stage the same power cells are presumed to be used as in the PA-stage. Therefore, only a few PA/DRV TGW ratios come into consideration, namely a PA/DRV TGW ratio of 4:4, 4:3, 4:2, and 4:1 as described in the beginning of this section. Since at a PA/DRV TGW ratio of 1:1, the driver-stage needs a lot of DC power which is not converted into output power, this PA/DRV TGW ratio is not very efficient. It delivers an improved gain performance at high compression levels but at the expense of a poor power-added efficiency. The trade-off between gain and PAE has always to be considered.

Relating to the results obtained from the simplified example at the beginning of this section, in this study PA/DRV TGW ratios of 4:2 and 4:1 are chosen to analyze the effect of a reduced driver stage with respect to the trade-off between the obtainable gain, PAE and bandwidth. The PA/DRV TGW ratio of 4:2 is convenient with most HPA designs found in literature (see Table 6.4) that commonly feature a PA/DRV TGW ratio between 2:1 and 3:1.

Table 6.4: AlGaN/GaN HEMT HPA MMIC designs at X-Band frequencies found in literature.

| **Cit.** | **Tech. + $l_g$ [$\mu$m]** | **Freq.-range [GHz]** | **$V_{DS}$ and chip size** | **PA/DRV GW ratio** | **pulsed $P_{out}$** | **PAE [%]** |
|---|---|---|---|---|---|---|
| [115] | MSL 0.2 | 9-11 | $V_{DS}$=30 V 2.2 mm$^2$ | 2.4/1 **=2.4/1** | 40-41.6 dBm =10-14.7 W | 23 |
| [23] | MSL 0.5 | 8.5-10 | $V_{DS}$=20 V 18.6 mm$^2$ | 8/4 **=2/1** | ≥43.2 dBm =21-28.5 W | 29-43 |
| [79] | MSL 0.25 | 8.5-9.5 | $V_{DS}$=25 V 18 mm$^2$ | 8.96/2.4 **=3.7/1** | 43-46 dBm =30-40 W | 33-38 |
| [101], [102] | MSL 0.25 | 8.7-11.5 | $V_{DS}$=30 V 12 mm$^2$ | 4/2 **=2/1** | 41.4-43 dBm =14-20 W | 31 |
| [126], [123] | CPW 0.3 | 10 | $V_{DS}$=35 V 13.5 mm$^2$ | 4/2 **=2/1** | 41.3 dBm =13.5 W | ≤25 |
| [125] | MSL 0.3 | 8-10 | $V_{DS}$=31 V 13.5 mm$^2$ | 4/2 **=2/1** | 39.5 dBm =8.9 W | ≥20 |
| [124] | MSL 0.3 | 8-11 | $V_{DS}$=40 V 13.5 mm$^2$ | 4.8/2 **=2.4/1** | 40-43 dBm =15-20 W | ≤25 |

In addition, according to Figure 6.6 and to [58], even a driver gate-width of only a quarter of the PA-stage gate-width is able to drive this PA-stage. The two resulting HPA designs are titled Duras (4:2) and Syrah (4:1), respectively. The specifications of both designs are summarized in Table 6.5. The last column of the table shows the PAE results of the last analytical assumption according to Equation 6.13 including the above specified net insertion losses.

Table 6.5: Design approaches of HPA MMICs Duras and Syrah.

| **MMIC** | **PA/DRV GW ratio** | **FET structure** | **Bandwidth** | **Theoretical limit according to Eq. 6.13** |
|---|---|---|---|---|
| **Duras** | 4:2 | 8×125 $\mu$m $l_g$=0.25 $\mu$m | 1.5-2 GHz | PAE=46.5 % |
| **Syrah** | 4:1 | 8×125 $\mu$m $l_g$=0.25 $\mu$m | 1.5-2 GHz | PAE=49.1 % |

On the one hand, the expected PAE of the HPA with large PA/DRV gate-width ratio is higher than the PAE of the HPA with the convenient PA/DRV TGW ratio of 2:1 but at the expense of smaller bandwidth and strongly reduced gain-reserve at the upper end of the frequency bandwidth. The two HPA MMICs Duras and Syrah were investigated with respect to power compression over bandwidth and PAE with regard to their different output/driver gate-width ratios of 4:2 and 4:1, respectively.

The determination of the DRV periphery depends not only on the required input power of the PA-stage transistors, but also on the frequency-dependent net insertion loss of the interstage matching network. Thereby, the ISMN in turn is dependent on the periphery of the DRV-stage. According to the number of FET cells forming the driver, additional combining networks become necessary within the interstage matching network leading to additional passive losses.

On the other hand, with the use of larger FET cells in the DRV, the same output power can be obtained with less transistors so that the number of transistor cells and the resulting number of combining networks in the interstage matching network can be reduced at the price of a lower power gain, especially at higher frequencies.

**OMN Specifications**

After the determination of the HPA periphery the output matching network has to be designed to transform the 50 Ω external load to $\Gamma_L$ for maximum output power of the transistor stage over a frequency range of about 2 GHz with a center frequency of $f_0$= 10 GHz. First, the optimum load impedance $\Gamma_{L,Pout}$ of the transistor stage is determined by load pull simulations of the HEMT cell. The main purpose of the OMN consists in this transformation with minimum net insertion losses according to Equation 6.15.
The structure of the OMN consists of three parts:

- a combining network for the four parallel FET cells used in the power-stage
- drain bias voltage supply path(s) to the drains of the PA-stage transistors
- an impedance transformation RF-path.

Since both designs consist of a PA-stage with four parallel-operated transistors, the OMN is the same for both MMICs. The number of inputs of the output matching network is equal to the number of power cells used in parallel in the PA-stage. Figure 6.9 shows the final structure of the layout of the output matching network together with the PA-stage transistors.

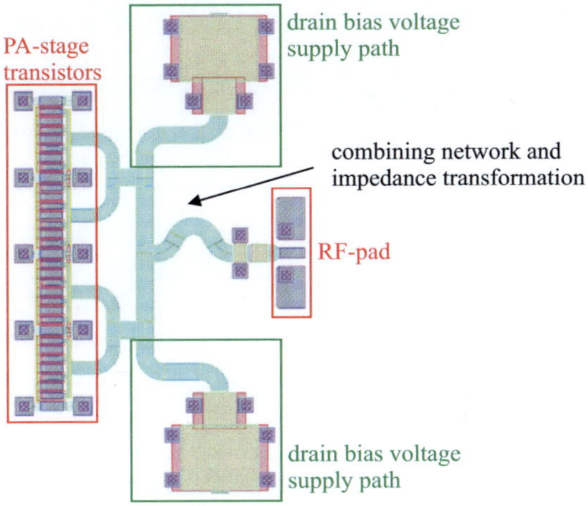


Figure 6.9: Layout of the output matching network and the PA-stage of both HPA MMICs.

In a first step the drains of the four PA-stage transistors are combined using microstrip lines and junction components. The lines are kept as short as possible to minimize additional losses but with respect to the conversion to a layout and the design rules, especially minimum distances between the single components such as lines and via holes. The combining network is done in a tree-like structure and kept symmetrical so that simulations can be reduced to a two-port simulation of just a part of the network by using mirror elements to reduce simulation time.

The bias stubs are realized with microstrip lines and large bias decoupling capacitors at the end of the bias stubs for DC blocking because the impedance level at the bias is unknown. One bias stub at each side of the network is used for symmetry reasons as described above and is dependent on the number of transistors used in parallel in the PA-stage and the consequential amount of DC current loading.

The third part of the network delivers the main impedance level transformation as well as the matching over the required bandwidth. At first, the external 50 Ω load is transformed into the optimum $\Gamma_L$ of the single transistor at $f_0$ as intended. In a further step, the matching over the required relative bandwidth of 20 % is obtained through optimization of the dimensions of the components used in the matching network. This part of the network includes a small series capacitor close to the output of the network (the RF-pad) that is relevant to provide DC decoupling to the output but is not intended for impedance transformation, as well as a single section lowpass network that consists of a capacitive reactance in form of a parallel capacitor shunting the higher load resistance along with a series resonant inductive reactance in form of microstrip line components. Additional lowpass matching networks would result in more complex multisection matching networks that deliver a matching over a broader bandwidth at the expense of higher net insertion losses. Since the aim of this matching network consists in the required impedance transformation over a 2 GHz bandwidth with minimum losses, only one lowpass impedance matching network is used.

In this network design the width of all microstrip lines is kept equal to a value of 92 $\mu$m which corresponds to 50 Ω. The most important issue for the selection of the line width is the amount of current that is expected to flow through the microstrip line. However, microstrip lines offer an inductive behavior that depends on the ratio of the length to the width of the line. Microstrip lines, especially in bias stubs, can be shortened by reducing their width with respect to the current capability.

Figure 6.10 depicts the simulated reflection coefficients $S_{11}$ and $\Gamma_{22}$ of the output matching network (left graph) as well as the calculated net insertion loss of the matching network $L_{OMN}$ (right graph). Figure 6.10(a) shows the output reflection coefficient $\Gamma_{22}$ (blue line) as well as the matching $S_{11}$ (orange line) to the simulated optimum output load impedance $\Gamma_{L,Pout}$ (red line) of the single power cell over the entire X-band frequency range from 8 GHz to 12 GHz in a 50 Ω environment. A matching of better than -18 dB is obtained over the required frequency range of 2 GHz from 9 GHz up to 11 GHz with an optimum match of better than -25 dB around the center frequency of 10 GHz. It can be seen that the output matching network is designed for best matching at the center frequency.

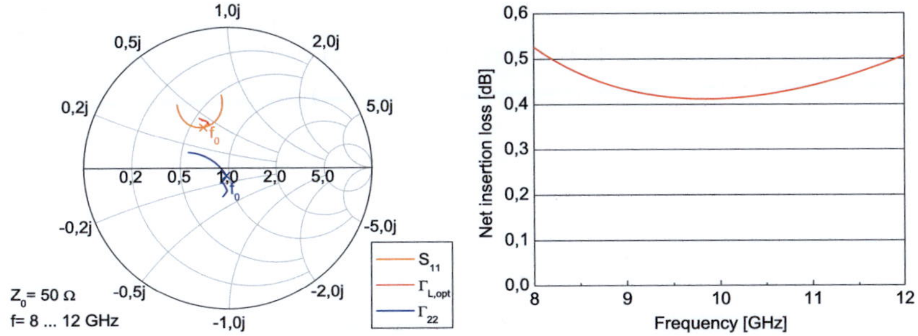


(a) Simulated reflection coefficients (b) Calculated net insertion loss

Figure 6.10: Simulated reflection coefficients $S_{11}$ and $\Gamma_{22}$ as well as calculated net insertion loss of the output matching network for the designs Syrah and Duras over the entire X-band frequency range.

The loss of the output matching network $L_{\mathrm{OMN}}$ primarly depends on the required bandwidth of the network, specified in the Q-factor that is defined as the ratio of the center frequency to the bandwidth BW [1], i.e. a high Q-factor implies a very small relative bandwidth.

$$Q = \frac{\omega_0}{BW} \tag{6.16}$$

Typical $L_{\mathrm{OMN}}$ values are in between 0.5-1 dB dependent on the bandwidth. According to Figure 6.10(b) this matching network delivers a very small loss of better than 0.5 dB for the entire X-band frequency range. Within the 20 % bandwidth the loss is even smaller than 0.44 dB. This low loss is realized by the use of as less elements as possible that are required to provide a good matching within the 2 GHz bandwidth.

### Critical Aspects to the ISMN

The structure of the interstage matching network primarily depends on the number of transistors used in parallel in the DRV- and PA-stage. In this chapter, two PA/DRV gate-width ratios are compared, a 4:2 (HPA Duras) and a 4:1 (HPA Syrah) ratio. Consequently, two different interstage matching networks are necessary for both HPA designs. Nevertheless, both ISMNs offer the same main functions, namely

- to transport the output power of the DRV-stage transistors to the input of the PA-stage transistors with minimum net insertion loss
- to provide bias voltage supplies to the drain of the DRV-stage transistors and separately to the gate of the PA-stage transistors.

In order to transport the output power of the DRV-stage transistors to the input of the PA-stage transistors with a minimum net insertion loss, the optimum load impedance for maximum output power of the transistors used in the DRV stage has to be transformed into the optimum input impedance for maximum output power of the PA-stage power cells. The impedance transformation between the two frequency dependent impedance levels including reactive parts makes the interstage matching network more challenging compared with the output matching network with a matching into a 50 $\Omega$ environment.

To realize the PA/DRV TGW ratio of 4:2, two power cells of the same size and periphery as in the output stage are used for the DRV-stage as well. The same power cells as in the PA-stage are used for the driver offering an optimum between gain and output power. The resulting interstage matching network of the HPA Duras (ratio 4:2) consists in a combining network of the two FETs used in parallel in the DRV-stage, followed by a drain bias stub on both sides of the network for drain biasing of the driver-stage transistors. After the drain bias stubs the network is split again into microstrip line and capacitor elements in form of a single section lowpass network for the impedance transformation as well as a gate bias stub on both sides for symmetrical gate biasing of the PA-stage transistors. Finally two more splitting networks to the input of the four PA-stage power cells are necessary. Within the gate bias paths inductors are used to realize high impedance RF blocking and resistors are included between the inductors and DC-blocking capacitors to stabilize the entire amplifier design.

The interstage matching network of the HPA Syrah (ratio 4:1) differs from the other one due to the fact that there is only one transistor used in the DRV-stage. In this case a splitting network is used first, followed by a drain bias stub on both sides of the network, a lowpass impedance transformation with the help of line and capacitor elements, two gate bias stubs, and finally two more splitting networks to the input of the four PA-stage power cells. The four bias stubs have a similar configuration as in the other interstage matching network.

Both ISMNs are constructed symmetrically for equal power supply of the PA-stage transistors, and to reduce the simulation complexity and time through the use of mirror elements as described in the OMN design paragraph. In this case, the several network configurations are designed with different line-capacitor element combinations. Out of results of the analyzed ISMN variants, the designs shown here are the best compromise between impedance transformation and low net insertion loss over the required BW. The layouts of both interstage matching networks are depicted in Figure 6.11.

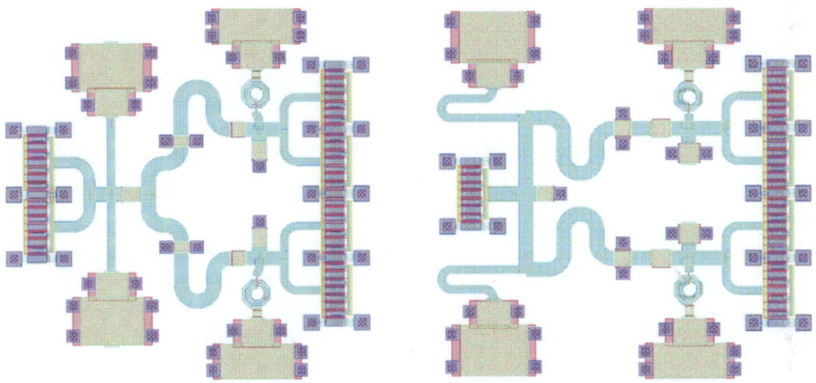

(a) ISMN of the HPA Duras (b) ISMN of the HPA Syrah

Figure 6.11: Layouts of the interstage matching networks and both transistor stages.

Since the matching over a specific frequency range is not ideal, the interstage matching networks are designed for better matching at the upper end of the frequency band than at the lower end of the frequency band because of the frequency-dependent sensitivity of the FET yielding to a sufficiently flat gain response.

Figure 6.12 gives a comparison between the simulated reflection coefficients $\Gamma_{11}$- and $\Gamma_{22}$ of the interstage matching networks for the designs Duras (on the left) and Syrah (on the right) over the entire X-band frequency range. Thereby, the center frequency of $f_0$= 10 GHz is marked in the figures. In both cases the optimum source and load impedance of the transistors used for maximum output power is depicted for the same frequency range. It can be seen that especially the $\Gamma_{11}$ curves are matched only in a small frequency range to the optimum output load impedance $\Gamma_{L,opt}$ (red line).

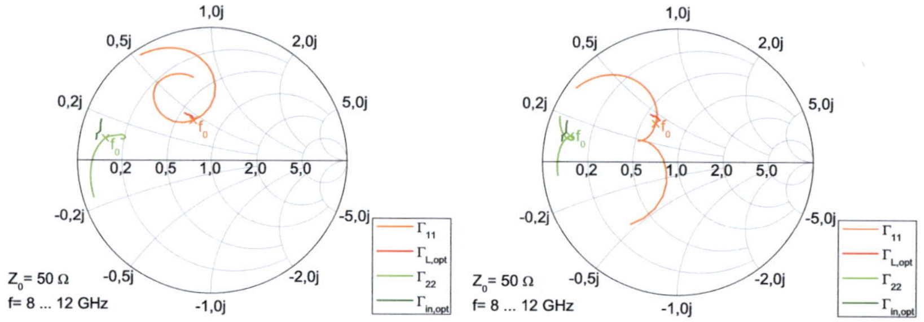


(a) ISMN of the HPA Duras (b) ISMN of the HPA Syrah

Figure 6.12: Simulated reflection coefficients $\Gamma_{11}$- and $\Gamma_{22}$ of the interstage matching networks for the designs Syrah and Duras over the entire X-band frequency range.

In Figure 6.13 the calculated net insertion loss of both interstage matching networks which is calculated for the ISMN using formula 3.20 is exhibited over the entire X-band frequency range. At the center frequency of 10 GHz both matching networks offer a minimum net insertion loss of approximately 1.5 dB. With the interstage matching network of the HPA Duras consisting of the smaller gate-width ratio of 4:2, the desired better matching at the upper end of the frequency band is realized with a constant net insertion loss of 1.5 dB. Otherwise, the ISMN of the HPA Syrah features a matching with such a low net insertion loss only for a small frequency range around 10 GHz. The net insertion loss is less than 2 dB for the required 20 % frequency range from 9 to 11 GHz with strongly increasing losses at the upper and lower ends of the frequency band. The required low-loss matching at the upper frequency band cannot be realized because of the difficult transformation of the relatively high PA / DRV gate-width ratio of 4:1.

As a result, the net insertion loss at the center frequency is almost the same for both matching networks while the PA / DRV TGW ratio has an impact on the low-loss matching over a broad bandwidth.

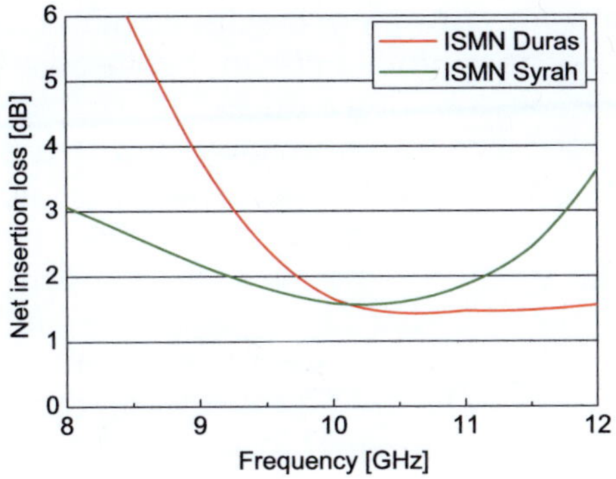


Figure 6.13: Calculated net insertion loss of the interstage matching networks for the designs Syrah and Duras over the entire X-band frequency range.

Consequently, the intentionally introduced higher losses of the ISMN towards lower frequencies to realize the desired gain flatness over the entire required frequency range is only possible with the interstage matching network of the HPA Duras.

### IMN Design Aspects

The last matching network that needs to be designed is the input matching network. The structure of this matching network is influenced by the number of driver stage transistors leading to the need of splitting networks to the single power cells associated with additional net insertion losses. The general requirements on the IMN include

- a splitting network to the single FET cells used in the driver stage
- a gate bias voltage supply path to the gates of the DRV-stage transistors
- an impedance transformation to match the transistors to the $50\,\Omega$ external load at the input of the amplifier and
- stabilization networks to provide external stability to the entire HPA.

The input matching network of the design Syrah with the 4:1 gate-width ratio requires no splitting network because the DRV-stage consists of only one transistor. This reduces the complexity of the input network. The impedance transformation to match the transistor to the required 50 Ω external load at the input of the amplifier is more challenging than in the OMN case corresponding to the much lower optimum input impedance of the transistor compared with the load impedance resulting in higher net losses. Different transformation networks with varying elements were tested with the resulting network topology shown in Figure 6.14 delivering the best results concerning best match and minimum losses.

Within both input matching networks a gate bias voltage supply path is added consisting of inductors, stabilization resistors and DC-blocking capacitors. At the input of the PA Syrah a series capacitance is used in the matching network for DC decoupling, which has a minor influence on the impedance transformation.

Another very important part of both input matching networks is the introduction of a parallel resistor-capacitance circuit to enable or enlarge the overall stability of the entire two-port amplifier at the price of increased network losses. Nevertheless, the input matching

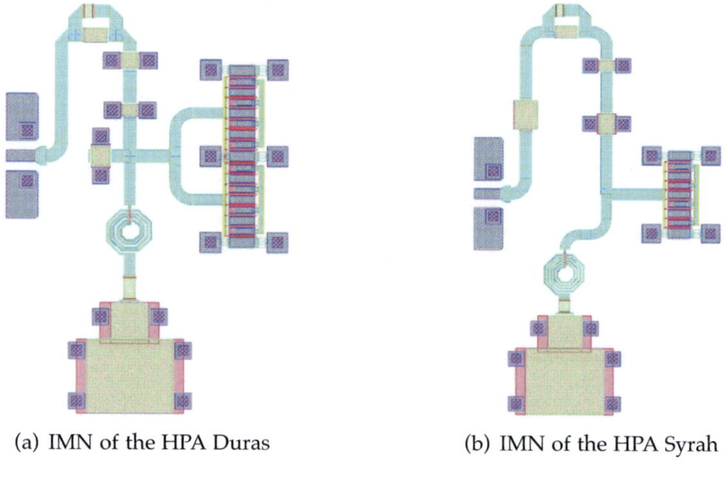

(a) IMN of the HPA Duras

(b) IMN of the HPA Syrah

Figure 6.14: Layouts of the input matching networks and driver-stages of the HPAs Duras and Syrah.

networks of both designs are designed for best matching near the upper end of the frequency band to compensate frequency dependent mismatch.

Figure 6.15 shows the simulated reflection coefficients $\Gamma_{11}$ and $S_{22}$ of both input matching networks over the entire X-band frequency

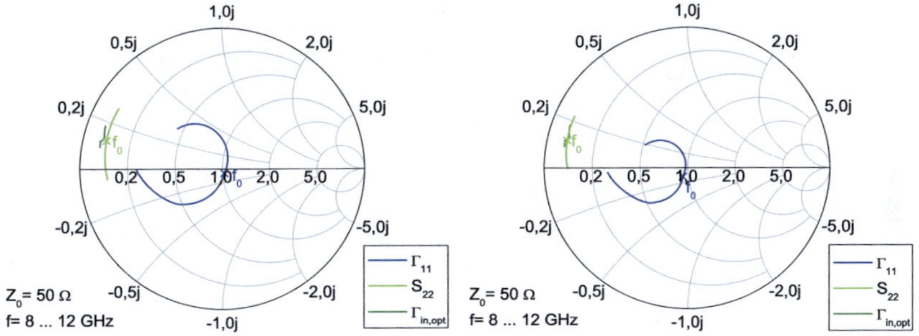


(a) IMN of the HPA Duras (b) IMN of the HPA Syrah

Figure 6.15: Simulated reflection coefficients $\Gamma_{11}$ and $S_{22}$ of the input matching networks for the designs Syrah and Duras as well as the optimum input impedance of the HEMT device, each over the entire X-band frequency range.

range as well as the optimum input impedance of the used transistor(s) for maximum output power within the same frequency range.

The simulated input reflection coefficient $\Gamma_{11}$ and the calculated net insertion loss of the input matching networks for the designs Syrah and Duras over the entire X-band frequency range are compared in Figure 6.16. Both matching networks offer a very good small-bandwidth matching of better than -25 dB around the center frequency of 10 GHz. Furthermore, the simulated net insertion loss at 10 GHz yields about 1.7 dB for both IMNs. As expected, the loss of the input matching network of the design Syrah using only one transistor in the driver stage is slightly lower than the other one at higher frequencies. The same accounts for the $\Gamma_{11}$ matching to the external 50 Ω that turns out to be better over the BW for the HPA Syrah (see Figure 6.15).

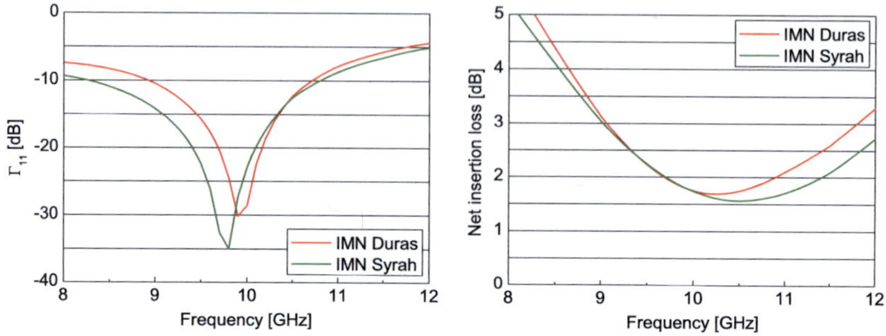


(a) Simulated input reflection coefficient $\Gamma_{11}$

(b) Calculated net insertion loss

Figure 6.16: Simulated input reflection coefficient $\Gamma_{11}$ and calculated net insertion loss of the input matching networks for the designs Syrah and Duras over the entire X-band frequency range.

### Compression Ratios of the DRV- and PA-Stage vs. Bandwidth

After the design of the matching networks which are described in detail above, a further important design step is to check the compression ratios of the DRV- and PA-stage over the desired bandwidth. To maximize the efficiency of the DRV stage, two options can be chosen: either the total cell size of this stage can be lowered, or the impedance of the load line can be increased. In this section, the cell size of the driver stage of the design Syrah is reduced compared with the design Duras because increasing the impedance level of the DRV output would result in a more critical matching transformation from the low-value real part of the PA-stage input impedance. The transformation to a double impedance level is more critical with respect to the quality factor or net insertion loss of the ISMN.

The PA / Driver gate-width ratios of the two HPAs Duras and Syrah were analyzed. In a first step, the simulated DRV, PA and total "intrinsic" power net gain of the HPAs Duras and Syrah were compared versus $P_{\text{in}}$ at $V_{\text{DS}}$= 30 V and $V_{\text{GS}}$= -2.25 V and at each center frequency, i.e. 10 GHz and 9.4 GHz, respectively. The results are shown in Figure 6.17. Looking at the gain compression of the HPA Duras in the figure on the left (Figure 6.17(a)), the DRV- and PA-stage are driven simultaneously into compression for low input power levels. For an input power of 28 dBm, the compression level of the PA-stage yields almost

7 dB while there is still only a small compression of the DRV-stage of 2.5 dB. Thus, the fact that the compression level of the PA-stage is almost three times higher than that of the driver implicates that a smaller driver would still be able to "drive" this PA-stage. In the figure on the right (Figure 6.17(b)), the DRV- and PA-stage "intrinsic" power gain curves of the HPA Syrah with a reduced driver gate-width are depicted for different input power levels. In this case, the resulting compression levels of both HPA transistor-stages are contrary to each other, compared with the HPA Duras. While the compression level of the PA-stage yields only 4 dB for an input power level of 28 dBm, the DRV-stage is driven into the double compression of 8 dB at the same input power level. Nevertheless, the obtained small-signal gain of the HPA Syrah is 2.5 dB higher than that of the HPA Duras, while both HPAs offer about the same power gain value for high input power levels.

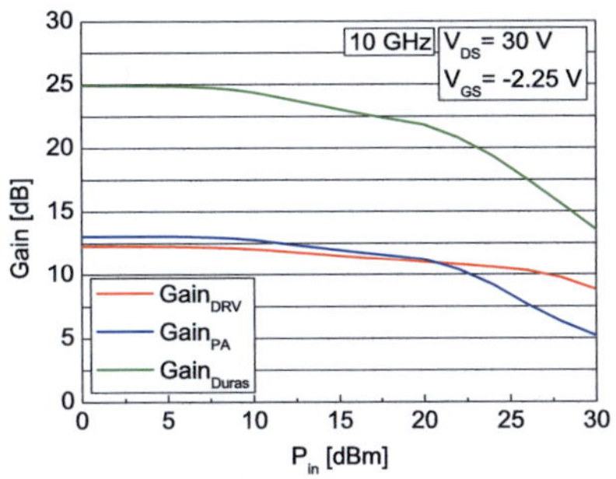


(a) HPA Duras at 10 GHz (b) HPA Syrah at 9.4 GHz

Figure 6.17: DRV-, PA- and total-"intrinsic"-power-net-gain of the MMICs Duras and Syrah simulated at $V_{DS}$= 30 V and $V_{GS}$= -2.25 V versus input power.

To check the compression ratio of the DRV- and PA-stage over the desired bandwidth, a combined frequency and power sweep is done in the harmonic-balance simulation of the entire HPA circuit [59]. Thereby, the "intrinsic" net gain of both stages (i.e. between gate and drain terminals of the DRV or PA FET cells) and the total "intrinsic" net gain between the DRV gates and the PA drains is plotted (Figure 6.18 and Figure 6.19). Each compression behavior is shown for an input power sweep from 0-28 dBm at $V_{DS}$= 30 V and $V_{GS}$= -2.25 V which is the small-signal bias point for class-AB operation. Both amplifiers are designed for a frequency range of 2 GHz from 9 up to 11 GHz with maximum gain and best matching between 9 and 10 GHz.

The DRV-, PA- and total-"intrinsic"-net-gain of the HPA Duras with a PA/DRV TGW ratio of 4:2 is shown in Figure 6.18. With regard to the increasing gain loss of the unit FET cell towards higher frequencies, primarily a detailed analysis of the gain compression of the HPA at the upper end of the required frequency bandwidth is essential. Looking at the HPA Duras, there is almost no compression (less than 2 dB) of the driver stage at the upper end of the frequency band at 11 GHz for an input power of 28 dBm, while the PA stage is driven into a relatively strong compression of approximately 8 dB at the same frequency and input power level. This means that this DRV stage has a reserve with which it could easily "drive" an even bigger PA stage. Consequently, a PA stage with a fixed gate-width of 4 mm

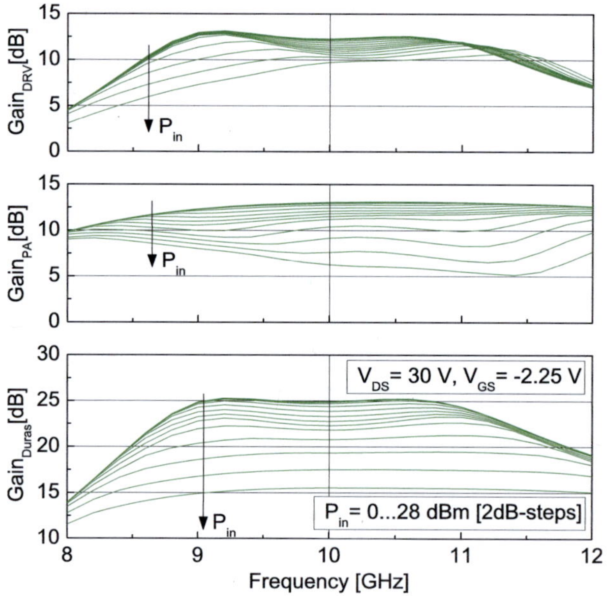


Figure 6.18: DRV-, PA- and total-"intrinsic"-net-gain of the MMIC Duras simulated at $V_{DS}$=30 V and $V_{GS}$=-2.25 V for different input power levels over the X-band frequency range.

could be driven by a smaller DRV-stage leading to the design Syrah with a PA/DRV TGW ratio of 4:1.

At the center frequency of 10 GHz compared with 11 GHz, as mark-ed in the figure, the compression level of the DRV-stage increases by about 1 dB as the compression level of the PA stage decreases by about the same value. Finally, at the lower end of the frequency band at 9 GHz, both compression levels are about the same. The resulting total "intrinsic" net gain compression of the entire MMIC Duras is found to be nearly frequency-independent over the entire frequency bandwidth from 9 up to 11 GHz.

Figure 6.19 displays the "intrinsic" net gain simulations of the HPA Syrah with a PA/DRV TGW ratio of 4:1. In this case, the driver-stage becomes "over-compressed" in comparison to the PA-stage at the upper end of the frequency band around 10.5-11 GHz so that the PA

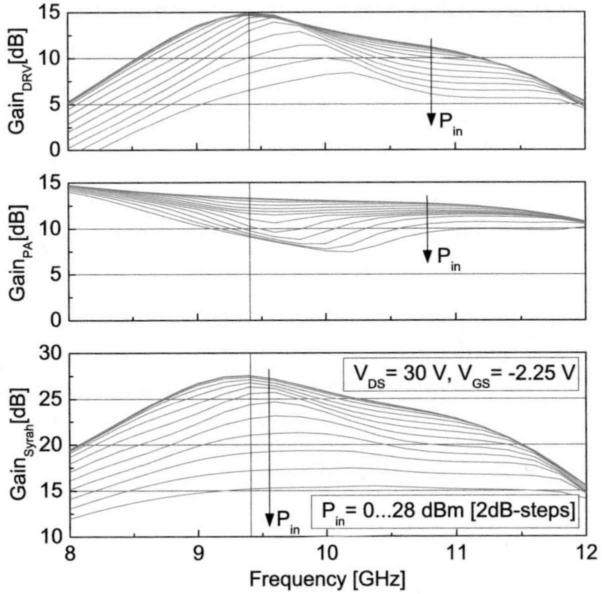


Figure 6.19: DRV-, PA- and total-"intrinsic"-net-gain of the MMIC Syrah simulated at $V_{DS}$= 30 V and $V_{GS}$= -2.25 V for different input power levels over the X-band frequency range.

driving power is at its limit with a 4:1 ratio for higher frequencies. At the center frequency of 10 GHz, the PA- and DRV-stage are simultaneously driven into compression, i.e. about 5 dB each. The compression level of the DRV stage increases with decreasing frequency and results in a 5 dB higher compression level than the PA stage at the lower end of the frequency band. The total "intrinsic" net gain of the HPA Syrah shows a maximum between 9 and 10 GHz at 9.4 GHz, but no frequency-independent gain compression over the entire frequency range such as the HPA Duras. In addition, the frequency range of the MMIC Syrah turns out to be smaller than that of the MMIC Duras with its smaller PA/DRV gate-width ratio.

This leads to the assumption that the amplifier Syrah with the PA/DRV TGW ratio of 4:1 is near its performance limit. Nevertheless, as already mentioned, the design Syrah yields an approximately 2.5 dB higher maximum small-signal-gain at the center frequency than the HPA Duras.

These considerations show that PA- and DRV-stage are driven simultaneously into compression for an optimized PA/DRV TGW ratio leading to a "softer" compression behavior of the entire HPA than in the case of an over-dimensioned driver ratio. For a broadband design with optimized PA/DRV TGW ratio, an acceptable "balance" ($\pm 2$ dB) of the DRV- and PA-compression level should be achieved over the entire bandwidth.

In conclusion, the small-signal and large-signal simulation results of both HPA MMIC designs are compared with each other and with the expected simplified analytical calculations from the beginning of this section. Table 6.6 summarizes all of the relevant data evaluated at 9.5 GHz, $P_{in}$= 28 dBm and $V_{DS}$= 30 V. Beginning with the small-signal behavior, the HPA Duras with the PA/DRV TGW ratio of 4:2 delivers a flat gain over a 2 GHz frequency bandwidth from 9 up to 11 GHz and holds a 3 dB bandwidth from 8.6-11.6 GHz, while the HPA Syrah with the enlarged PA/DRV TGW ratio of 4:1 shows a gain peaking between 9 GHz and 10 GHz with its maximum value at 9.4 GHz. In addition, the achievable 3 dB bandwidth of the MMIC Syrah is reduced to 1.8 GHz. However, the maximum obtained small-signal gain is about 2.5 dB higher than at the HPA Duras. As expected, the higher maximum small-signal gain comes at the prize of a reduced bandwidth. With regard to the large-signal simulations at above mentioned conditions, both designs offer the same saturated output power and power gain of about 20 W and 15 dB, respectively.

At this high gain-compression level, the maximum PAE values are reached, namely 42 % with the HPA Duras and 44 % with the MMIC Syrah. According to the simplified calculations, these simulation results show the same tendency of PAE improvement based on an enlarged PA / DRV TGW ratio. Nevertheless, the values of the simulated PAE enhancement is not as high as expected, i.e. about 4.5-5 % below the calculated PAE value. This is the fact due to the simplified assumptions of the analytical calculations that are valid only for a single-frequency, while these HPA MMICs are designed for a frequency bandwidth of approximately 2-3 GHz.

Table 6.6: Comparison of the small-signal- and large-signal-simulation results of the HPA MMICs Duras and Syrah evaluated at 9.5 GHz, $P_{in}$= 28 dBm and $V_{DS}$= 30 V as well as the expected PAE on the basis of the simplified analytical calculations at comparable conditions.

| **MMIC** | **PA/DRV GW ratio** | **Freq.-range + 3 dB BW** | **max. small-signal gain + curve char.** | **LS-Sim. of PAE** | **PAE after Eq. 6.13** |
|---|---|---|---|---|---|
| **Duras** | 4:2 | 8.6-11.6 GHz<br>3 GHz | 25 dB<br>flat gain compression | 42 % | 46.5 % |
| **Syrah** | 4:1 | 8.7-10.5 GHz<br>1.8 GHz | 27.5 dB<br>gain peak at 9.4 GHz | 44 % | 49.1 % |

### Stabilization Issues

After the design of all matching networks, the stability of the entire amplifier as well as the stability of the individual transistor stages must be analyzed. External instabilities such as the instabilities of the complete amplifier including bias oscillations at low frequencies and RF oscillations close to or even within the frequency bandwidth of the amplifier need to be located and eliminated with the help of $\mu$-factor analysis as described in detail in Section 3.3.1. The HPA Syrah with its high PA / DRV TGW ratio offering more small-signal gain over a small bandwidth is more difficult to stabilize because of the extreme gain behavior. As a result, the resistor values in the gate-bias paths of the ISMN need to be doubled compared with the HPA Duras to enable external stability. Within the input matching network, it is reverse. Within the IMN gate-bias path, the HPA Syrah needs only half the resistance compared with the design Duras to achieve stability. This is due to the fact that the matching of only one transistor input to the $50\,\Omega$ environment is much easier because of the higher impedance level than the matching of 2 parallel operated transistors.

Besides the determination of external stability, the internal stability of the individual stages is examined by a Z-parameter analysis at the input and output of each transistor as described by Ohtomo [72] (see Section 3.3.1). In both designs, odd-mode resistors are included between the inputs and outputs of all parallel operated transistors of each stage to enhance the stability of the used FETs.

### HPA Chip Layouts

Figure 6.20(a) and 6.20(b) shows the chip image of the dual-stage HPA Duras and Syrah, respectively. The chip size is standardized to $4.5 \times 3\,\text{mm}^2$ for both designs. In the figures the varying driver stages are marked with red ellipses to highlight the different cell sizes of the driver stages.

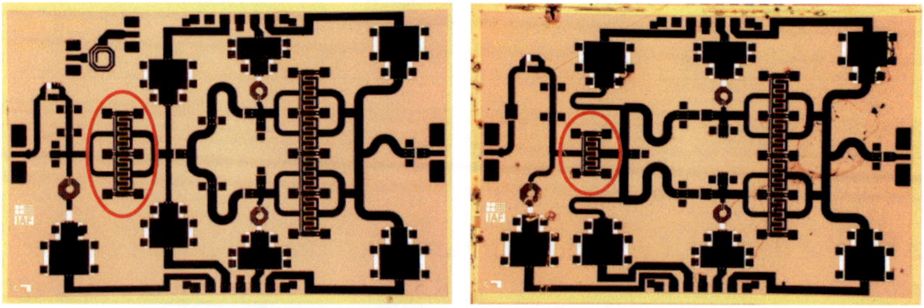

(a) HPA Duras (4:2)

(b) HPA Syrah (4:1)

Figure 6.20: Photographs of the dual-stage HPAs Duras and Syrah with the varying driver-stage marked. The chip size is $4.5 \times 3\,\text{mm}^2$ for both designs.

### 6.1.4 PA/DRV TGW Relating HPA Performances

Although a large number of wafers is available, the data shown below is taken from the same wafer to reduce any variations.

Initially, the S-parameters of both HPAs Duras and Syrah were measured in pulsed mode at $V_{DS}$= 30 V and $I_{DS}$= 150 mA/mm. The results of the measured S-parameters are depicted in Figure 6.21 emphasizing the small-signal gain $S_{21}$ and the $S_{11}$ and $S_{22}$ matching of both HPAs. The 3 dB frequency bandwidth of the HPA Duras from 8.5 up to 10.5 GHz is marked in green in the figure. The design Duras offers a flat maximum gain of approximately 22.5 dB from 8.7-10 GHz, while the HPA Syrah with the reduced PA/DRV gate-width ratio delivers an slightly improved maximum gain of 23.1 dB at 10 GHz and a reduced 3 dB bandwidth from 8.9-10.4 GHz compared with the HPA Duras. Both designs deliver a decrease in the measured maximum gain performance of -2.5 dB (HPA Duras) and -4.4 dB (HPA Syrah) compared to the simulations. While the MMIC Syrah offers almost the same 3 dB bandwidth as predicted by the simulations, the measured frequency range of the HPA Duras is shifted 0.5 GHz downwards on the frequency axis. Furthermore, both designs show their best input matching between 9 and 10 GHz of better than -10 dB and a GaN specific poor output matching for all frequencies.

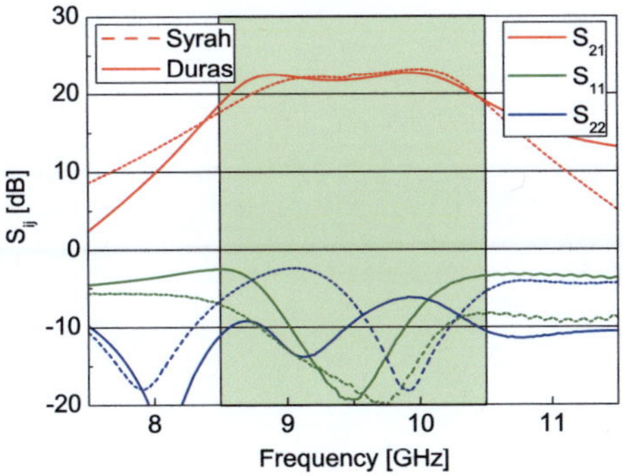


Figure 6.21: Pulsed-S-parameters of the dual-stage MMICs Duras (solid line) and Syrah (dashed line) measured at $V_{DS}$= 30 V and $I_{DS}$= 150 mA/mm.

All large-signal measurements are performed in pulsed-RF mode with a duty cycle of 10 % and a pulse length of 100 $\mu$s. Figure 6.22 illustrates a comparison of the broadband large-signal simulations (line) and pulsed-RF measurements (line with symbols) of the dual-stage HPAs Duras and Syrah, respectively. Power, gain, and PAE results of both amplifiers are depicted at $V_{DS}$= 30 V and for an input power of $P_{in}$= 26 dBm.

For the HPA Duras (Fig. 6.22(a)), simulated and measured data are in good agreement in the intended frequency range up to 10.5 GHz, however, there are some deviations for the upper X-band range from 10.5 GHz on. The green area marks the bandwidth from about 8.4 GHz up to 10.6 GHz of the MMIC Duras with a measured PAE of $\geq$ 30 %, an output power of more than 42 dBm, and an associated constant gain of 15 dB.

The amplifier Syrah (Fig. 6.22(b)) shows deviations for simulation and measurement results over the entire X-band frequencies, where the simulated data turn out to be too optimistic compared with the measured ones. Nevertheless, the principal trend is the same. The MMIC Syrah exhibits 1.6 GHz bandwidth from 9.1 - 10.7 GHz showing almost similar results as for the HPA Duras. Especially the inferior bandwidth behavior of the amplifier Syrah corresponds to the challenging PA / DRV gate-width ratio of 4:1.

By comparing both broadband measurement results, it can be

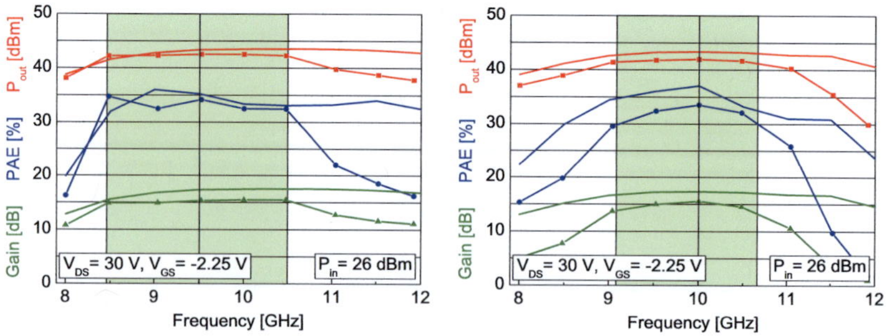


(a) HPA Duras

(b) HPA Syrah

Figure 6.22: Broadband large-signal simulations (line without symbols) and pulsed-RF measurements (line with symbols) of the dual-stage HPAs Duras and Syrah at $V_{DS}$= 30 V.

seen, that the decrease of the driver stage does not yield in a PAE enhancement as desired, but has a critical impact on the overall bandwidth. The reason could be the "over-compressed" DRV-stage compared with the PA-stage of the MMIC Syrah with the 4:1 PA/DRV TGW ratio.

Figure 6.23(a) displays the pulsed RF-measurements of the HPA MMIC Duras measured at 9.5 GHz versus input power. The biasing of the power-sweep measurements is the same compared with that of the frequency-sweep measurements. For easy comparison with the broadband measurement results, the data of this graph is marked at the same compression level of -5 dB at the same input power of 26 dBm. The achieved output power at this compression level is 42.5 dBm, i.e. $\approx$ 17.8 W, with a PAE of 36 % and an associated gain of 16.4 dB. The saturated output power of the MMIC Duras with this biasing reaches $P_{\text{sat}}$= 43 dBm (20 W) with an associated $PAE_{\text{max}} > 40\,\%$. These results fit perfectly to the broadband measurement results, i.e. the obtained output power is the same, the power gain is about 1 dB higher, and the achieved PAE turns out to be 2 % higher.

The pulsed-RF-measurements of the HPA Syrah are depicted over input power in Figure 6.23(b). The power sweep is shown for a drain voltage of $V_{\text{DS}}$= 30 V with a slightly adjusted gate biasing of -2 V compared with $V_{\text{DS}}$= -2.25 V for the frequency-sweep measurements. The frequency for the power-sweep is chosen to be 10 GHz

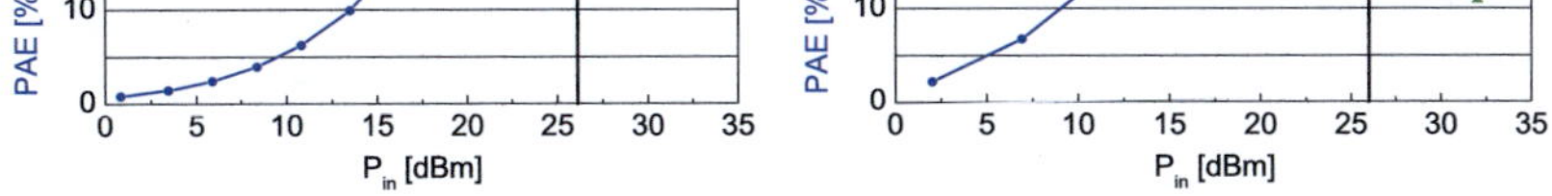


(a) HPA Duras

(b) HPA Syrah

Figure 6.23: Pulsed-RF measurement results over input power of the HPAs Duras and Syrah.

which turned out to be the optimum frequency for maximum output power, PAE, and power gain according to the broadband pulsed-RF-measurements from Figure 6.22(b). Once more, the results at 26 dBm input power are marked for comparison with the broadband measurements, indicating a -8 dB compression level. At this high compression level, an output power of 42 dBm (15.8 W) with an associated gain of 15.9 dB and a PAE of 37 % is reached. Again, power- and frequency-sweep measurement results are in good agreement. The output power is nearly identical, and the power gain and PAE are slightly better measured with the power-sweep, i.e. 0.9 dB and 4 %, respectively.

### Conclusions

Table 6.7 gives a comparison of the large-signal simulation and pulsed-RF-measure-ment results of the two HPA MMICs Duras and Syrah. Both the broadband and the power-sweep results are evaluated at the same input power level of 26 dBm and a drain voltage of 30 V.

On closer examination of the broadband simulations and measurements, the HPA Duras shows a very good agreement of simulation and measurement results of the obtained PAE, output power and power gain values within the frequency-range of 8.4-10.6 GHz, where the PAE is higher than 30 %. The maximum simulated and measured PAE is approximately 35 %. However, the simulations of the MMIC Syrah are quite too optimistic compared with the measurements. Even though, the simulations of the HPA Syrah offer an improvement of PAE with similar output power and associated power gain compared with the MMIC Duras, the measurement results deviate from the simulations showing an inferior behavior. Nevertheless, the principal trend is the same. For frequencies up to 11 GHz the measured PAE turns out to be about 5 % below the expected and simulated value. Consequently, the obtained measured maximum PAE yields only 34 % that is 1 % less than the maximum PAE of the MMIC Duras by offering about the same power gain and an output power reduced by approximately 0.7 dB. This result stands in contradiction to the expected PAE improvement due to a higher PA/DRV gate-width ratio according to the theory presented at the beginning of this section and the simulation results.

Compared to the broadband measurements, the pulsed-RF power sweeps show slightly improved results for the obtained output power,

Table 6.7: Comparison of the large-signal-simulation and measurement results of the HPA MMICs Duras and Syrah evaluated at $P_{in}$=26 dBm and at $V_{DS}$=30 V.

| MMIC | Broadband Simulations | Broadband Measurements | Power-Sweep Measurements |
|---|---|---|---|
| | at $V_{DS}$=30 V, $V_{GS}$=-2.25 V | | |
| | 8.4-10.6 GHz (2.2 GHz) | | at 9.6 GHz |
| **Duras 4:2** | $PAE \geq 30\,\%$<br>$PAE_{max} = 35.5\,\%$ | $PAE \geq 30\,\%$<br>$PAE_{max} = 35\,\%$ | $PAE = 36\,\%$<br>$PAE_{max} \geq 40\,\%$<br>at $P_{in} \geq 30$ dBm |
| | $P_{out} \geq 43$ dBm | $P_{out} \geq 42.2$ dBm | $P_{out} = 42.5$ dBm<br>$= 17.8$ W |
| | | $P_{out,max} = 42.7$ dBm | $P_{out,sat} = 43$ dBm<br>at $P_{in} \geq 30$ dBm |
| | $G_p = 15\text{-}17$ dB | $G_p = 15$ dB | $G_p = 16.4$ dB |
| | at $V_{GS}$=-2.25 V | | at $V_{GS}$=-2 V |
| | 9.1-10.7 GHz (1.6 GHz) | | at 9.8 GHz |
| **Syrah 4:1** | $PAE \geq 33\,\%$<br>$PAE_{max} = 37\,\%$ | $PAE \geq 30\,\%$<br>$PAE_{max} = 34\,\%$ | $PAE = 37\,\%$<br>$PAE_{max} = 41.3\,\%$<br>at $P_{in} \geq 30$ dBm |
| | $P_{out} \geq 42.7$ dBm | $P_{out} \geq 41.4$ dBm | $P_{out} = 42$ dBm<br>$= 15.8$ W |
| | | $P_{out,max} = 42$ dBm | $P_{out,sat} = 42.9$ dBm<br>at $P_{in} \geq 30$ dBm |
| | $G_p = 17$ dB | $G_p = 14\text{-}16$ dB | $G_p = 15.9$ dB |

power gain, and PAE. Here, the HPA Syrah with the adjusted biasing indicates the better PAE values (approximately 1 % above that of the HPA Duras) as expected from the theory and HPA simulations.

Furthermore, the design goal of a saturated output power of $P_{sat} \approx 20$ W with a PAE target of 40% was nearly reached with both designs. Nevertheless, the additional design target for a 2 GHz bandwidth within X-band frequencies could only be obtained for the HPA Duras. The higher gate-width ratio of 4:1 of the MMIC Syrah leads to strongly increasing net insertion losses in the interstage matching network towards the upper end of the frequency band which lead to a lower bandwidth reserve, especially at higher frequencies. In addition, the ISMN of the HPA Syrah offers a stronger mismatch according to Figure 6.12 that also causes a limitation of bandwidth.

As a result, the higher gate-width ratio of 4:1 seems to be "overcompressed" relating to the measured data. On the one hand, the desired and expected PAE improvement of a HPA with a PA/DRV TGW ratio of 4:1 can only be reasonably verified by the power-sweep measurements, and, on the other hand, the power amplifier Syrah shows a quite reduced operating bandwidth (1.6 GHz) whereas the PAE is higher than 30 % compared with the HPA Duras with a corresponding bandwidth of 2.2 GHz.

### 6.1.5 Optimized PA/DRV TGW Ratio

Based on the results obtained from the comparison of the HPA MMICs Duras and Syrah, an improved HPA MMIC with optimized PA/driver gate-width ratio titled HPA Sacy was developed. The PA/DRV TGW ratio is chosen in between the gate-width ratios of the MMICs Duras (2:1) and Syrah (4:1) leading to a new PA/DRV TGW ratio of 3:1.

In addition to the improved PA/driver gate-width ratio with the purpose of PAE enhancement, this MMIC had also the target specifications of high output power levels $\geq 20$ W and high drain operating voltages up to 35 V. According to the two designs Duras and Syrah, the MMIC Sacy was also presumed to reach its target performance over a 1.5-2 GHz bandwidth within X-band frequency range. The center frequency of the HPA Sacy was designed to be 9 GHz.

#### Design Aspects

In a first design step, the total gate-width of the output and driver stage were chosen in accordance with Figure 6.8 from the last section and according to the design goals specified above. For the HPA Duras with a total output gate-width of 4 mm a saturated output power of 20 W could just be achieved over a very narrow bandwidth around the center frequency. However, this design comprises a design goal of $\geq 20$ W output power over a 2 GHz frequency bandwidth. Therefore, a 6 mm gate-width periphery was chosen in the output stage of this MMIC design due to the requirement of higher output power over a large frequency bandwidth realized on the basis of power scaling. Again, four power cells are used in parallel in the PA-stage. The resulting single HEMT cell yields a 1.5 mm gate periphery. An optimum gain performance over the required X-band frequencies is achieved once a $10\times150\,\mu$m gate geometry is used as a best compromise between gate-finger count and finger width. Figure 6.24 depicts the comparison between the simulated MSG/MAG-curves of the two GaN HEMT variants with 1 mm (red curves) and 1.5 mm (green curves) gate geometry at $V_{DS}$= 30 V and at two different gate voltages of $V_{GS}$= -2 V and $V_{GS}$= -2.25 V. It has to be considered that the larger FET cell ($10\times150\,\mu$m) features a MSG-MAG intersection point at lower frequencies rather close to 10 GHz compared with the 1 mm FET cell with an $8\times125\,\mu$m gate geometry, dependent on the bias point. Consequently, the gain of the larger FET cell is significantly reduced for high frequencies.

The resulting output matching network of the HPA Sacy is designed in the style of the OMN of the amplifiers Duras and Syrah taking the modified $\Gamma_L$ of the larger FET cell into account. Again, the main focus of the OMN is a smallest possible net insertion loss to reach maximum output power. The loss of the output matching network persists below 0.65 dB for all relevant frequencies up to 10 GHz.

Since the PA/DRV gate-width ratio is set to 3:1, the resulting driver-stage has a total gate-width of 2 mm. As for the design Duras, two 1 mm FET cells with an $8\times125\,\mu$m gate geometry are used for the driver-stage. ISMN and IMN are designed on the basis of the design Duras but optimized for maximum output power and PAE in

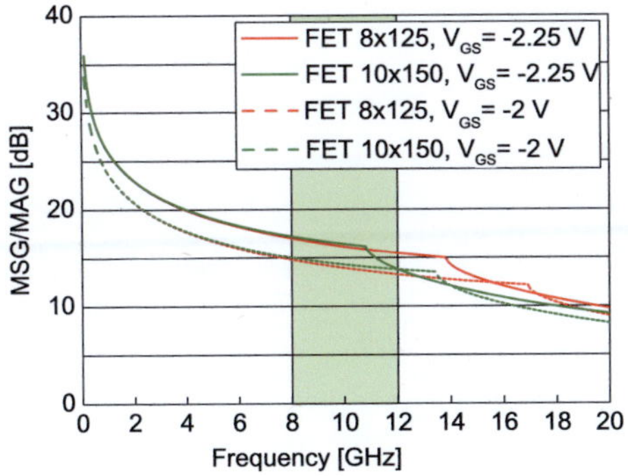


Figure 6.24: Comparison of the simulated MSG/MAG-curves of the two GaN HEMT variants with 1 mm (red curves) and 1.5 mm (green curves) gate geometry at $V_{DS}$= 30 V and at two different gate voltages of $V_{GS}$= -2 V and $V_{GS}$= -2.25 V.

consideration of the modified FET cell in the output stage and the adapted PA/DRV gate-width ratio.

**Compression behavior of the DRV- and PA-stage**

Following, the simulated gain compression behaviors of the DRV- and PA-stage as well as of the entire HPA Sacy were analyzed. At first, the DRV-, PA- and total-"intrinsic" net gain of the MMIC Sacy was simulated at $V_{DS}$=30 V and $V_{GS}$=-2.5 V versus input power. In Figure 6.25 the gain compression curves are shown for the center frequency of 9.4 GHz. Compared with the gain curves of the HPAs Duras and Syrah (figure 6.17), the resulting compression levels of the DRV- and PA-stage of the MMIC Sacy are more homogenous, even though the small signal gain of the PA-stage is 2.5 dB higher than that of the DRV-stage. This is mainly due to the different load of the transistors used for the driver- and PA-stage at 9.4 GHz that causes a not perfectly matched ISMN. In addition, the HEMT with the 10×150 $\mu$m gate-geometry delivers a slightly higher MSG value of approximately 0.4 dB at the frequency of 9.4 GHz as can be seen in Figure 6.24.

While the PA-stage transistors start going into compression at an input power of 15 dBm, the DRV-stage FETs go constantly into compression starting at small input power levels. For an input power of

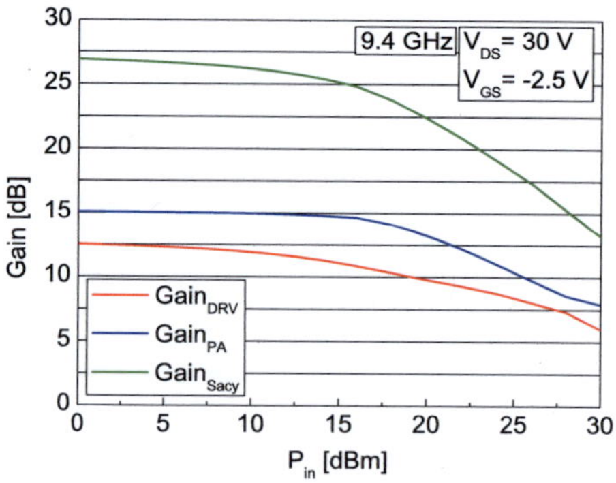


Figure 6.25: DRV-, PA- and total-"intrinsic"-net-gain of the MMIC Sacy simulated at $V_{DS}$=30 V and $V_{GS}$=-2.5 V at 9.4 GHz versus input power.

28 dBm, the compression levels of the DRV- and PA-stage yield 5.2 dB and 6.5 dB, respectively. Thus, the desired "balance" (±2 dB) of the DRV- and PA-compression level is reached with the PA/DRV TGW ratio of 3:1.

Figure 6.26 depicts the DRV-, PA- and total-"intrinsic" net gain compression of the MMIC Sacy simulated at a biasing of $V_{DS}$=30 V and $V_{GS}$=-2.5 V and for the same input power levels as chosen in the last section for the HPAs Duras and Syrah for easy comparison.

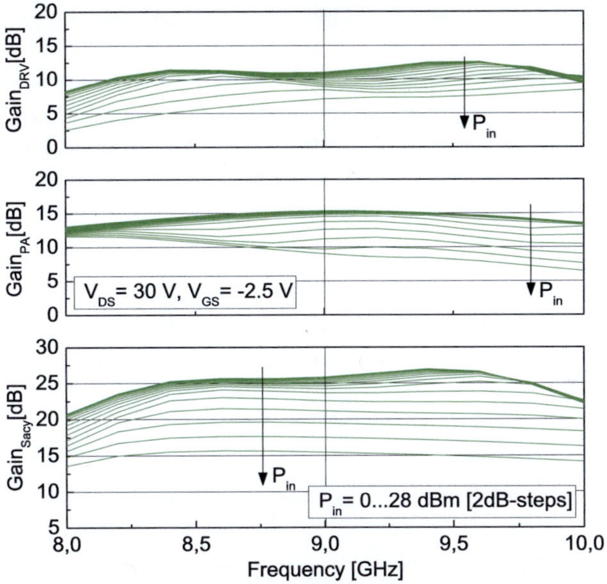


Figure 6.26: DRV-, PA- and total-"intrinsic"-net-gain of the MMIC Sacy simulated at $V_{DS}$=30 V and $V_{GS}$=-2.5 V for different input power levels over the X-band frequency range.

On closer examination of the compression levels of DRV- and PA-stage at the upper (9.8 GHz) and lower (8.3 GHz) ends of the frequency band as well as for the center frequency of 9 GHz, an overall compression level in between the two other designs Duras and Syrah can be obtained as expected. At the upper end of the frequency band the compression level of the PA-stage with about 7 dB corresponds to the double compression level of the DRV-stage. The resulting difference between DRV and PA compression levels at the upper end of the frequency band is about 3.5 dB. Since the PA compression level decreases with decreasing frequencies down to 4 dB at the lower end of the frequency band, and the DRV compression level increases respectively up to 6 dB at the lower end of the frequency band, a "balanced" ($\pm 2$ dB) compression level of DRV and PA is achieved at least over the lower half of the bandwidth of the entire HPA design Sacy.

The simulated "intrinsic" net gain of the MMIC Sacy delivers an almost linear gain curve above 25 dB over the entire frequency range from 8.3 GHz up to 9.8 GHz with a maximum gain of nearly 27 dB at 9.5 GHz. In addition, the entire HPA shows a "flat" compression level over the whole bandwidth of about 1.5 GHz. Compared with the other two designs Duras and Syrah, the MMIC Sacy takes advantage of both: it reaches the same maximum small-signal gain and the same bandwidth as the HPA Syrah but features the desired flat compression behavior as the HPA Duras.

**HPA chip layout**

The MMIC chip layout of the dual-stage HPA Sacy is displayed in Figure 6.27. Even though the PA-stage size is larger than that in the other two MMICs, the chip size is still the same as for the HPA Duras and Syrah, namely $4.5 \times 3\,\mathrm{mm}^2$.

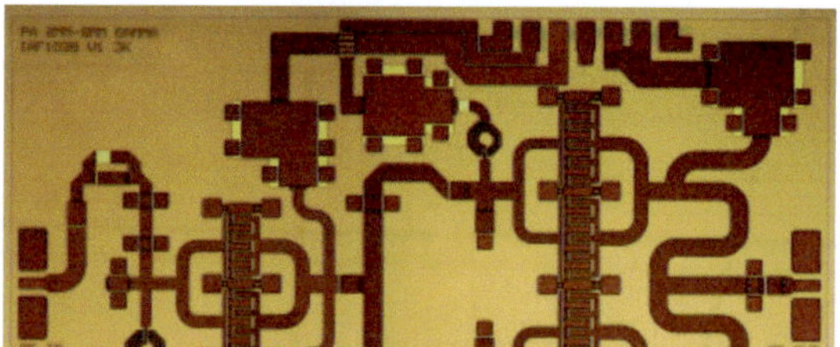

Figure 6.27: Photograph of the dual-stage HPA Sacy. The chip size is $4.5 \times 3\,\mathrm{mm}^2$.

**Experimental Results**

The large-signal performance of the HPA Sacy was obtained on a wafer with an epitaxial III-nitride growth such as the other two designs Duras and Syrah. All frequency and power sweeps were measured in pulsed-RF mode in a broadband system with an input and output load of 50 Ω.

The measured pulsed small-signal S-parameters of the HPA Sacy are depicted in Figure 6.28. The bias condition was chosen to be $V_{DS}$= 30 V with $I_{DS}$= 120 mA. A small-signal maximum gain of 24.5 dB could be obtained at a frequency of 9 GHz with an input and output matching of better than -5 dB as marked in the figure. There is a 2.5 dB decrease of the measured maximum gain compared to the simulation results. Further, the 3 dB frequency bandwidth from 8.2-9.4 GHz is marked in green in the figure signifying almost the same but slightly reduced bandwidth as predicted by the simulations.

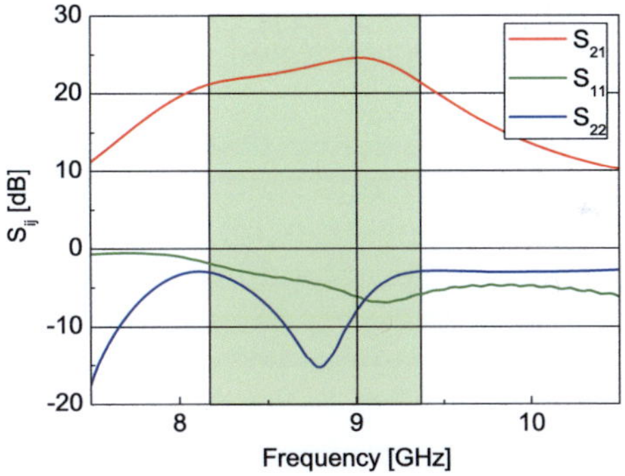


Figure 6.28: Pulsed-S-parameter measurements of the HPA Sacy taken at $V_{DS}$= 30 V and $I_{DS}$= 120 mA.

Figure 6.29(a) depicts the pulsed-RF frequency sweep of the MMIC performed with a constant input power of $P_{in}$= 24 dBm for a 4 dB compression level at $V_{DS}$= 32 V. Further measurement results of the HPA Sacy can be found in [57].

According to the broadband measurements of the HPA Duras and Syrah, the area highlighted in green shows again the bandwidth in which the HPA delivers an output power greater than 40 dBm with an associated PAE of $\geq$ 30 %. The measured bandwidth covers and

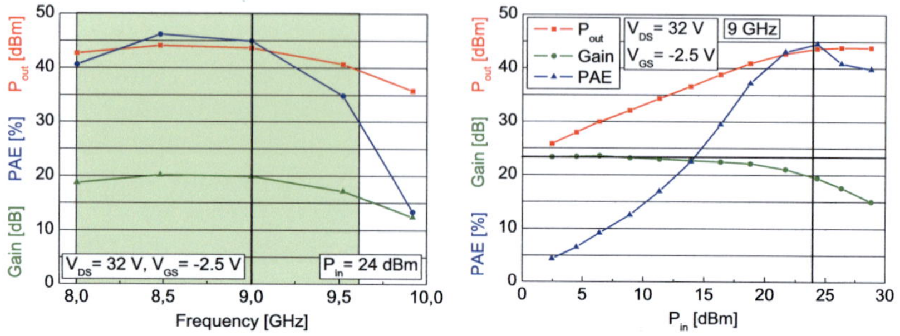


(a) Frequency sweep at $P_{\text{in}}$=24 dBm (b) Power sweep at 9 GHz

Figure 6.29: Pulsed-RF-measurements of the HPA Sacy at $V_{\text{DS}}$=32 V.

even exceeds the intended and designed bandwidth from 8.3 GHz up to 9.7 GHz. Referring to the design goals, an output power of $\geq$ 20 W is required. This design target is achieved for a frequency range from 8 GHz up to 9.2 GHz with an associated PAE and gain of 40-46 % and 18.5-20 dB, respectively.

Even though a maximum output power of 44 dBm (25 W) with an associated maximum PAE of 46 % can be obtained at a frequency of 8.5 GHz, the figure on the right (6.29(b)) shows a power sweep of the same MMIC with the same bias condition at the center frequency of 9 GHz. At 24 dBm input power (4 dB gain compression) an output power of 43.7 dBm ($\approx$ 23.5 W) can be reached together with 44.5 % PAE and 19.5 dB power gain.

As can be seen in the figures above, the design goal of $P_{\text{out,sat}}$=20 W with an associated PAE$\geq$ 40 % can be obtained with the HPA MMIC Sacy. Furthermore, the HPA Sacy with its sophisticated PA/DRV gate-width ratio of 3:1 being in between those of the MMICs Duras and Syrah delivers a considerably PAE enhancement of 2-4 %, compared with the two other designs with the same bandwidth as the design Syrah.

**Conclusions**

With all three HPA designs the design goals of $P_{\text{out,sat}}$= 20 W with an associated PAE above 40 % could be achieved. Nevertheless, the impact on the PA/DRV gate-width ratio on the PAE and output power performance is shown in this section. Table 6.8 contains the main pulsed-RF measurement results of all three HPAs (Duras, Syrah and Sacy) for comparison. In this case, the HPAs Duras and Syrah were performed on the same wafer, while the HPA Sacy was evaluated with a posterior run containing an improved technology relating to e.g. a reduced low-frequency dispersion of the HEMTs. Nevertheless, the results are comparable.

As expected from the simulations, the linear gain of a HPA becomes larger with increasing PA / DRV gate-width ratio. Nevertheless, this statement correlates with the limitation of the obtained bandwidth of the single HPA designs. The higher linear gain values are only achieved at the expense of smaller bandwidth.

The same assumption is made for the power-added efficiency according to the analytical calculations. However, the principle trend that the PAE increases with increasing PA / DRV gate-width ratio is only valid when the HPAs Sacy (3:1) and Syrah (4:1) are compared with the HPA Duras (2:1). When comparing the HPA Sacy with the HPA Syrah, no further PAE enhancement is achieved. In contrast, the PAE becomes lower again. This is due to the fact that the ISMN matching becomes more difficult with increasing PA / DRV TGW ratio leading to higher mismatches and net insertion losses, especially towards higher frequencies, and therefore to a limited frequency bandwidth and a lower PAE. In addition, with increasing PA / DRV TGW ratio, the DRV compression level grows higher at comparable input power near output saturation with the effect of a lower DRV-stage power gain that also causes a decrease in the PAE of the entire HPA.

As a result, in this comparison the PAE value becomes higher for HPAs with a PA / DRV TGW ratio of up to 3:1, but lower again for HPA designs with larger PA / DRV TGW ratios than 3:1 because of an "over-compression" behavior (HPA Syrah). In addition, the designs with higher PA / DRV TGW ratios become more narrow in bandwidth because of the higher mismatch in the interstage matching networks with increasing PA / DRV TGW ratio.

Finally, the saturated output power mainly depends on the gate-width periphery of the PA-stage. As expected, since the HPAs Duras and Syrah feature the same PA-stage periphery of 4 mm, both HPAs deliver a saturated output power of approximately 20 W, even though the obtained output power of the HPA Syrah is slightly reduced compared with that of the HPA Duras. Furthermore, with the HPA Sacy with the larger PA-stage periphery of 6 mm, theoretically 30 W of saturated output power are expected. However, only 24.5 W are realized. The reason for the reduced output power is the relatively high PAE that is obtained with the HPA Sacy compared to the other two designs. In contrast to the HPA MMICs Duras and Syrah, the HPA Sacy was designed for maximum PAE rather than maximum $P_{out}$.

Table 6.8: Comparison of the pulsed-RF measurement results of the HPA MMICs Duras, Syrah, and Sacy.

| **Main features** | **Duras 2:1** | **Syrah 4:1** | **Sacy 3:1** | **Evaluation** |
|---|---|---|---|---|
| lin. Gain | 22 dB | 24 dB | 23.5 dB | increases with increasing PA / DRV TGW ratio |
| $PAE_{max}$ | 40 % | 41.3 % | 45 % | expected to increase with increasing PA / DRV TGW ratio, but HPA Syrah is "overcompressed" |
| BW with PAE$\geq$30 % | 8.4-10.6 GHz (2.2 GHz) | 9.1-10.7 GHz (1.6 GHz) | 8-9.6 GHz (1.6 GHz) | |
| $P_{out,sat}$ | 43 dBm = 20 W | 42.9 dBm = 19.5 W | 43.9 dBm = 24.5 W | $\approx$ 20 W with 4 mm PA-stage of HPAs Duras and Syrah $\approx$ 25 W with 6 mm PA-stage of HPA Sacy |

## 6.2 Harmonic Termination Effects on PAE Behavior

This section explains how the PAE of a power amplifier from Class-AB up to Class-B condition can be increased by terminating the even harmonics. In this case, the largest harmonic besides the fundamental is the second harmonic.

The influence of the harmonics on a transistor can be demonstrated by determining all components of the output current waveform of a transistor. Thereby, the DC supply provides the mean component that is decreasing monotonically as the conduction angle is reduced (see Section 3.1.2). The magnitude of the DC supply, the fundamental current component as well as all higher harmonic components can be calculated according to [25] with the help of Fourier analysis. The current waveform yields

$$I_{\mathrm{n}} = \frac{1}{\pi} \cdot \int_{-\alpha/2}^{\alpha/2} \frac{I_{\mathrm{max}}}{1 - cos(\alpha/2)} \left[cos\theta - cos(\alpha/2)\right] cos(n\theta) d\theta \ . \qquad (6.17)$$

The results from Equation 6.17 for n=0 (DC supply) up to n=5 ($5^{th}$ harmonic) are shown in Figure 6.30. By examining the curves, it can be seen that the DC component decreases as expected. In addition, throughout Class-AB range the only significant harmonic besides the fundamental is the second. All higher harmonics show amplitudes close to zero for Class-A up to Class-B operation.

It is clear that the transistor output current contains some significant harmonic components, especially second harmonic in Class-AB operation. Harmonic termination (HT) however, presents a method of controlling these harmonic components presented to a transistor at the input and output, principally by means of wave shaping of the voltage and current waveforms at the FET to reduce the overall power dissipated in the transistor [53]. As a result, the increase in efficiency arises from the reduced transistor power dissipation. The effective power that is delivered at the harmonics above the fundamental (especially at the second harmonic in Class-AB operation) is prevented due to a total reflection of the even harmonics at the load. Thus, this reflected power at the harmonics is added to the RF fundamental output power leading to an enhancement in efficiency.

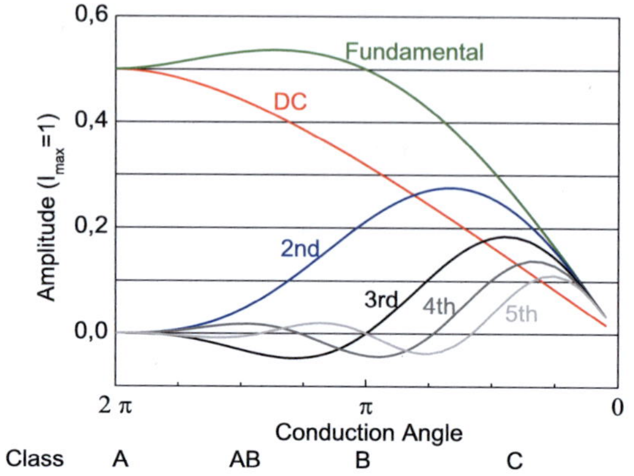


Figure 6.30: Fourier analysis of the current waveform components vs. conduction angle after [25].

According to the Fourier analysis of the current waveform (6.17), the first three components of the drain current (the DC, fundamental, and $2^{nd}$ harmonic component) result in

$$I_{DS}(\text{Class-A}) = \frac{I_{max}}{2} + \frac{I_{max}}{2} \cdot \cos(\omega_0 t) + 0 \cdot \cos(2\omega_0 t) \quad (6.18)$$

$$I_{DS}(\text{Class-B}) = \frac{I_{max}}{\pi} + \frac{I_{max}}{2} \cdot \cos(\omega_0 t) + \frac{2}{3\pi} I_{max} \cdot \cos(2\omega_0 t) \quad (6.19)$$

for Class-A and Class-B operation, respectively. As can be seen in Figure 6.30 and Equations 6.18 and 6.19, the second harmonic component is zero in Class-A operation but increases towards Class-B operation up to $2/(3\pi)I_{max}$, i.e. the harmonic current ratio of the $2^{nd}$ harmonic to the fundamental is $I_2/I_1 = 0.42$. Thus, the output power at the second harmonic results in $P_{out,2nd} = (0.42)^2 \cdot P_{out,1st}$. Therefore, second harmonic termination in Class-B operation theoretically leads to a PAE enhancement of

$$PAE_{withHT} = (1 + (0.42)^2) \cdot PAE_{noHT} = 1.176 \cdot PAE_{noHT} \,. \quad (6.20)$$

Harmonic shorts connected directly to the input or output of a device are used to prevent any harmonic voltage from being generated so that $V_{DS}$ is a sinewave whose magnitude can be set by the load resistor value to generate maximum permissible voltage swing [25]. Consequently, a short circuit at even harmonic frequencies is a promising option to improve the PAE of a HPA as well.

Initially the harmonic termination can be realized using short-circuit stub-networks that are implemented in a shunt connection into the amplifier design directly at the reference planes of the transistor device. Any shift of the HT position relating to these reference planes makes the short-networks sensitive to varying source/load impedances [113]. Figure 6.31 shows the circuit block diagram of a single-stage HPA with second harmonic short-circuits at the input and output of the intrinsic device, as well as input and output matching networks that are still required to match the active device at the fundamental frequency.

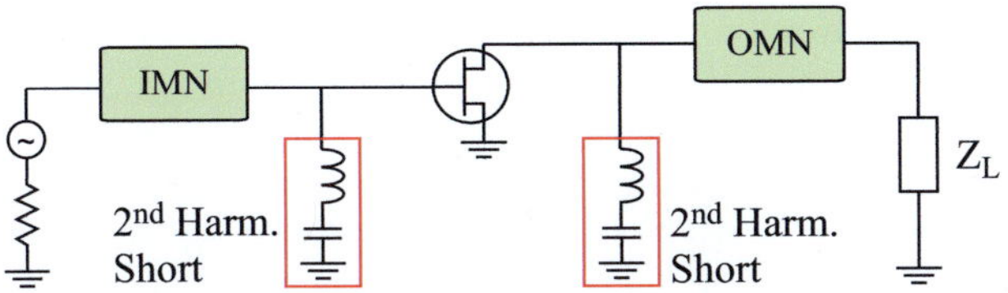


Figure 6.31: Circuit block diagram of a single-ended HPA with second harmonic short-circuits.

In this study, the effect of harmonic termination on the PAE performance is demonstrated at device level first by using harmonic balance simulations with a real device large-signal model as described in Section 3.1.3. Next, the short-circuit stub-structures are designed separately as discussed in detail in Subsection 6.2.2. Finally, the $2^{nd}$ harmonic stubs are implemented into a single-stage HPA design at the input and/or output of the device to determine the influence of the second harmonic short-stubs concerning the PAE performance on MMIC design level.

### 6.2.1 Influence of Harmonics on the PAE of a HEMT Device

First, the influence of the second harmonic on device level is analyzed. For this purpose, an $8\times125\,\mu$m AlGaN/GaN T-gate HEMT device without source-connected shields, an aluminum concentration of the AlGaN layer of 22 %, and a gate-to-gate pitch of $50\,\mu$m is characterized at 10 GHz in Class-AB operation on the basis of a nonlinear large-signal model to determine the optimum load and input impedances for maximum PAE values at the fundamental operating frequency with and without second harmonic tuning.

In a first step, fundamental load-pull (LP) simulations of an AlGaN/GaN HEMT are performed at the fundamental frequency for which the second harmonic is set to $50\,\Omega$ load. In the next step, additional load-pull simulations of the same transistor are done to determine the maximum PAE for optimum second harmonic tuning.

#### AlGaN/GaN HEMT without Consideration of Harmonics

Load-pull simulations of the device are performed for Class-AB condition ($V_{DS}$= 30 V and $V_{GS}$= -2.25 V) and for an input power level of $P_{in}$= 32 dBm (4 dB gain compression) at 10 GHz. This LP-simulation means that the complex output load $\Gamma_L$ is varied to sweep the Smith chart on a rectangular (real+imaginary) grid for the evaluation of the optimum load location for maximum $P_{out}$ and PAE at the fundamental. Thereby, the second harmonic load is set to $50\,\Omega$. The resulting optimum $\Gamma_S$ and $\Gamma_L$ for maximum output power and for maximum PAE are shown in the Figures 6.32(a) and 6.32(b), respectively. In addition, the source and load stabilization circles are depicted in the figures (green curves) exhibiting a stable region at the outside of each circle. Comparing the $\Gamma_L$ values for maximum output power and PAE, their locations in the Smith chart strongly differ from each other. Thus, a maximum PAE in a power amplifier design can only be obtained by terminating the HEMT device with $\Gamma_{L,PAE}$ at the expense of lower output power and vice versa.

The resulting PAE values of the HEMT device terminated with $\Gamma_{L,PAE}$ are depicted in Figure 6.33 as a function of the real- and imaginary-part of the output load $\Gamma_L$ in 0.05 Re$\{\Gamma_L\}$ and Im$\{\Gamma_L\}$ steps from -0.7 to 0.7. As can be seen in the figure, a maximum PAE of 54.2 % can be obtained for $\Gamma_L$ = -0.1 + j0.65.

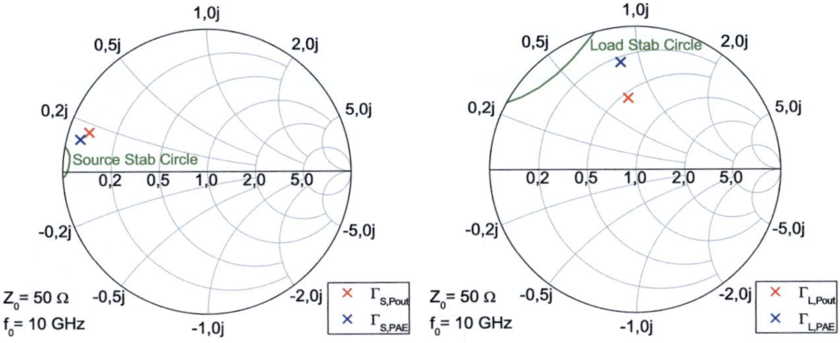


Figure 6.32: Simulated optimum $\Gamma_S$ and $\Gamma_L$ for maximum output power and PAE of an AlGaN/GaN HEMT at 10 GHz, $P_{in}$= 32 dBm, and biased at $V_{DS}$= 30 V and $V_{GS}$= -2.25 V.

Table 6.9 gives a comparison of the simulated output power, PAE, and power gain performance for a 1 mm FET cell terminated with $\Gamma_{L,PAE}$ for maximum possible power-added efficiency and with $\Gamma_{L,Pout}$ for optimum output power values. There is an improvement of 4 % in PAE at the expense of 0.5 W of output power by terminating the device with $\Gamma_{L,PAE}$.

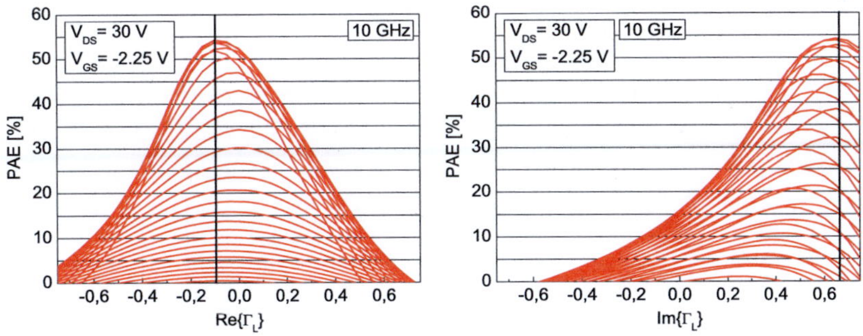


Figure 6.33: The simulated PAE of an AlGaN/GaN power cell with $W_g$= 1 mm is depicted versus the real and imaginary part of $\Gamma_L$ based on LP-simulations and performed at 10 GHz and $P_{in}$= 32 dBm. Bias point: $V_{DS}$= 30 V and $V_{GS}$= -2.25 V.

Table 6.9: Simulated $P_{out}$, PAE, and gain values for a 1 mm AlGaN/GaN HEMT terminated with the optimum load impedance for maximum $P_{out}$ and PAE at $V_{DS}$=30 V, $P_{in}$=32 dBm, and 10 GHz.

| **Load, $\Gamma_L$** | **Source, $\Gamma_S$** | $P_{out}$ [W/mm] | **PAE** [%] | **Gain** [dB] |
|---|---|---|---|---|
| $\Gamma_{L,Pout}$= 0.5 ∠95.7° = (27.6+j36.9)Ω | $\Gamma_{S,Pout}$= 0.86 ∠161.4° = (3.9+j8.1)Ω | 4.6 | 50.2 | 10.4 |
| $\Gamma_{L,PAE}$= 0.66 ∠98.8° = (17.4+j39.8)Ω | $\Gamma_{S,PAE}$= 0.89 ∠163.6° = (3.0+j7.2)Ω | 4.1 | 54.2 | 10.8 |

Finally, Figure 6.34 shows the simulated power-sweep of the FET device terminated with the above obtained optimum $\Gamma_{L,PAE}$ at the same bias condition and frequency. Thereby, the maximum PAE value of 54.2 % is reached for an input power level of 32 dBm at 4 dB gain compression. The output power at the same compression level yields 36.1 dBm with an associated gain of 10.8 dB.

A further PAE enhancement is possible according to the theoretical analysis from the beginning of this section with the help of second harmonic termination. Based on Equation 6.20 and wrongly assuming a Class-B operation, the simulated maximum PAE of 54.2 % can

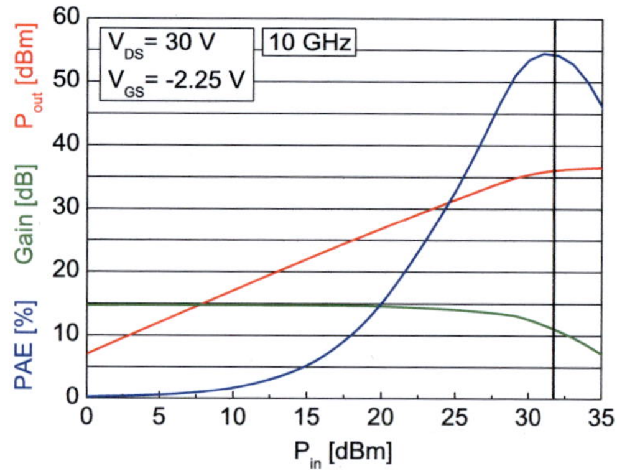


Figure 6.34: Simulated power-sweep of a 1 mm AlGaN/GaN HEMT at 10 GHz terminated with the opt. $\Gamma_{L,PAE}$ at $V_{DS}$=30 V and $V_{GS}$=-2.25 V.

be increased up to $(1 + (0.42)^2) \cdot 54.2\,\% = 63.8\,\%$, i.e. a further PAE improvement of 9.6 %. Taking into account that this transistor is operated in Class-AB operation, the expected PAE enhancement due to second harmonic tuning is slightly less than the theoretically calculated value.

### AlGaN/GaN HEMT in Consideration of Second Harmonic Tuning

Additional load-pull simulations are performed regarding the second harmonic tuning to determine optimum harmonic source and load impedances.

Figure 6.35 shows the simulated PAE versus the magnitude (from 0 to 1 in 0.05-steps) and phase (from 0° to 360° in 5-degree steps) of the output $2^{nd}$ harmonic termination of a 1 mm AlGaN/GaN power cell. Thereby, the HEMT device is terminated with $\Gamma_{L,PAE}$ at the fundamental frequency of 10 GHz and operated at an input power level of 32 dBm and a biasing of $V_{DS}$= 30 V and $V_{GS}$= -2.25 V. According to the figure, the maximum obtained PAE of 54.2 % with the $2^{nd}$ harmonic set to 50 Ω can be improved by more than 7 % by tuning the second harmonic over a specific but small angle. The maximum PAE value yields 61.5 % and can be realized with an optimized second harmonic termination with a magnitude of 1 (lossless case) and a phase of 120°. As can be seen in the figure, for a value of zero for

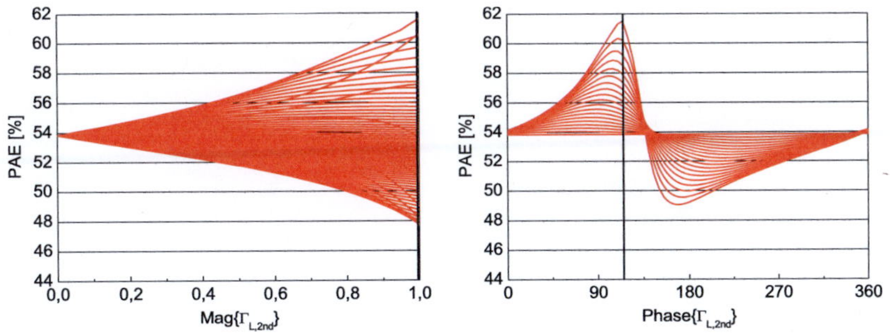


Figure 6.35: Simulated PAE versus magnitude and phase of the output $2^{nd}$ harmonic termination of a 1 mm AlGaN/GaN HEMT at 10 GHz and $P_{in}$= 32 dBm under consideration of the optimum $\Gamma_{L,PAE}$ for maximum PAE at $f_0$. Bias point: $V_{DS}$= 30 V and $V_{GS}$= -2.25 V.

Mag$\{\Gamma_{L,2nd}\}$ a PAE of 54.2% is obtained, indicating the initial case for which the $2^{nd}$ harmonic is set to 50 Ω.

To provide a second harmonic short at $2f_0$ at the output of the HEMT device, a short-circuit stub is used that can be designed only with a phase of 180° (see Subsection 6.2.2). To realize the short at the optimum simulated phase of 120°, the short-circuit stub-network is not implemented directly to the output reference plane of the HEMT device, but has to be 60° phase-shifted.

Figure 6.36 depicts the simulated power-sweep of the 1 mm AlGaN/GaN HEMT terminated with the optimum $\Gamma_{L,PAE}$ at the fundamental frequency and an optimum second harmonic termination at the output of the device (lines without symbols) as well as the second harmonic set to 50 Ω load (lines with symbols) for comparison. As can be seen in the figure, the gain characteristic is completely unaffected by the termination of the second harmonic, while the achievable output power yields only 3.6 W/mm at an input power level of 32 dBm compared with 4.1 W/mm in the case with the $2^{nd}$ harmonic set to 50 Ω load. Nevertheless, at higher input power levels the same output power is obtained independent of the termination of the second harmonic. The comparable PAE at 32 dBm input power amounts

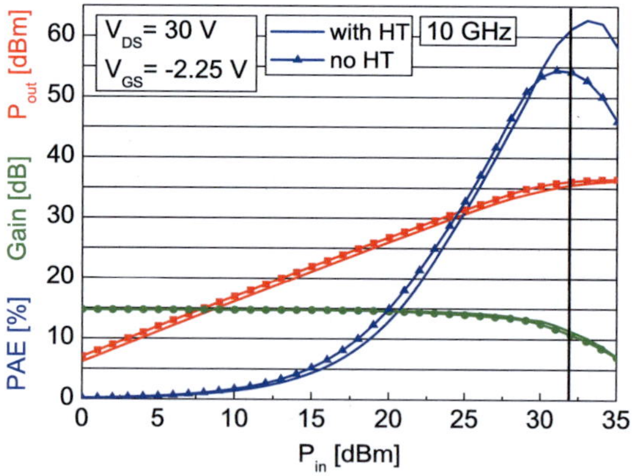


Figure 6.36: Simulated power-sweep of the 1 mm AlGaN/GaN HEMT at 10 GHz terminated with $\Gamma_{L,PAE}$ at the fundamental frequency and an optimum output second HT as well as the second harmonic set to 50 Ω load for comparison. Bias point: $V_{DS}$= 30 V and $V_{GS}$= -2.25 V.

61.3 %. However, a maximum PAE of 62.7 % is also obtained at a higher input power level of 33 dBm.

Compared with the output 2nd harmonic termination, an equivalent magnitude and phase sweep of the second harmonic is done for the input side of the HEMT device at the same conditions as before. The PAE values that are obtained in this way are depicted in Figure 6.37, showing a maximum PAE of 61 % at a magnitude of one and a phase of 180°. This corresponds directly to the short location in the Smith chart, i.e. an input short-circuit stub needs to be inserted directly in front of the gate of the HEMT device.

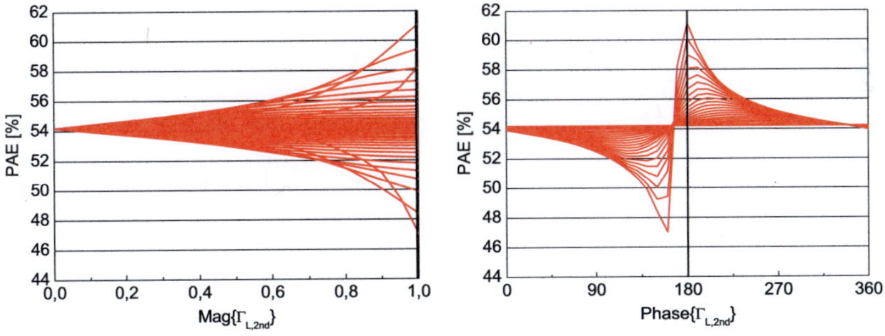


Figure 6.37: Simulated magnitude and phase of the $2^{nd}$ harmonic at the gate side of an AlGaN/GaN power cell with $W_g$= 1 mm at 10 GHz and $P_{in}$= 32 dBm under consideration of the optimum $\Gamma_{L,PAE}$ for maximum PAE. Bias point: $V_{DS}$= 30 V and $V_{GS}$= -2.25 V.

Table 6.10 summarizes the PAE performance of a 1 mm FET cell terminated with the optimum load impedance $\Gamma_{L,PAE}$ at the fundamental frequency of 10 GHz and ideal second harmonic terminations at the input and/or output of the device operated at a biasing of $V_{DS}$= 30 V and $P_{in}$= 32 dBm. In the first case, the second harmonic is set to 50 Ω leading to a PAE of 54.2 %. With an optimum $2^{nd}$ harmonic termination at the drain side of the HEMT the PAE yields 61.3 %, i.e. an absolute PAE enhancement of 7.1 %. An ideal $2^{nd}$ harmonic tuning at the gate side of the device delivers a comparable PAE improvement of 6.8 % with a maximum PAE value of 61 %. Using both, an optimized input and output harmonic termination, a further PAE improvement can be obtained with a maximum PAE of 62.3 %, yielding a PAE enhancement of 8.1 %. The output power and power gain

Table 6.10: Simulated PAE values for a 1 mm FET cell terminated with the optimum load $\Gamma_{L,PAE}$ for maximum PAE at $V_{DS}$=30 V, $P_{in}$=32 dBm, and 10 GHz and in consideration of an optimum lossless $2^{nd}$ harmonic tuning at the input and/or output of the device.

| $2^{nd}$ harm. at output [magnitude/phase] | $2^{nd}$ harm. at input [magnitude/phase] | PAE | Δ PAE |
|---|---|---|---|
| 0 | 0 | 54.2 % | - |
| 1 ∠120° | 0 | 61.3 % | +7.1 % |
| 0 | 1 ∠180° | 61.0 % | +6.8 % |
| 1 ∠120° | 1 ∠180° | 62.3 % | +8.1 % |

values stay nearly constant for all four cases.

According to the simulated load-pull results above, second harmonic tuning delivers an absolute PAE improvement of approximately 7 % in Class-AB operation, independent of the $2^{nd}$ harmonic termination at the input or output of the AlGaN/GaN HEMT device in the above mentioned operation mode and with the assumption of a lossless $2^{nd}$ harmonic tuning. This result fits together with the above theoretical PAE enhancement of 9.6 % in Class-B operation.

In reality, harmonic tuning is realized with the help of short-circuit stubs that are implemented close to the input or output of the HEMT device leading to some non-negligible losses. Taking these losses into account, the achievable PAE improvement becomes drastically reduced. Table 6.11 gives an overview of the effective PAE enhancement due to input and output harmonic tuning compared with the case with an optimized load $\Gamma_{L,PAE}$ at the fundamental and with all higher harmonics terminated to 50 Ω. For example, if the magnitude of the second harmonic termination yields only 0.8, the PAE improvement decreases from the theoretical possible value of approximately 7 % down to 3 %, independent of the $2^{nd}$ harmonic tuning at the gate or drain side of the HEMT device as can be directly derived from the diagrams 6.35 and 6.37, respectively. According to Table 6.11, the influence of input and output second harmonic termination is comparable to a slightly better performance of the harmonic termination at the load.

Figure 6.38 shows the simulated time domain output voltage and current waveforms at the intrinsic device level with an optimized

Table 6.11: Influence of a lossy $2^{nd}$ harmonic tuning at the input or output of an AlGaN/GaN HEMT device on the PAE performance. The FET cell is terminated with the optimum load $\Gamma_{L,PAE}$ at the fundamental, at $V_{DS}$= 30 V, and $P_{in}$= 32 dBm.

| $2^{nd}$ harm. at input [mag./phase] | PAE | Δ PAE | $2^{nd}$ harm. at output [mag./phase] | PAE | Δ PAE |
|---|---|---|---|---|---|
| 0 | 54.2 % | - | 0 | 54.2 % | - |
| 1.00 ∠180° | 61.0 % | +6.8 % | 1.00 ∠120° | 61.3 % | +7.1 % |
| 0.95 ∠180° | 60.0 % | +5.8 % | 0.95 ∠120° | 60.2 % | +6.0 % |
| 0.90 ∠180° | 58.9 % | +4.7 % | 0.90 ∠120° | 59.0 % | +4.8 % |
| 0.85 ∠180° | 58.0 % | +3.8 % | 0.85 ∠120° | 58.0 % | +3.8 % |
| 0.80 ∠180° | 57.3 % | +3.1 % | 0.80 ∠120° | 57.2 % | +3.0 % |

$\Gamma_{L,PAE}$ at the fundamental and with all higher harmonics terminated to 50 Ω (green lines), as well as with an optimized load at $f_0$ and an optimized input (blue lines) and output (red lines) second harmonic tuning, respectively. All simulations are performed at 10 GHz, $P_{in}$= 32 dBm, and a biasing of $V_{DS}$= 30 V and $V_{GS}$= -2.25 V.

(a) no HT compared with output HT (b) no HT compared with input HT

Figure 6.38: Simulated time domain output voltage (solid lines) and current (dashed lines) waveforms at the intrinsic device level with optimized load at the fundamental and harmonic loads of 50 Ω and optimized input/output second harmonic terminations, respectively. All simulations are performed at 10 GHz, $P_{in}$= 32 dBm, and a biasing of $V_{DS}$= 30 V and $V_{GS}$= -2.25 V.

The figure on the left (Figure 6.38(a)) illustrates the effect of output second harmonic termination. According to Figure 6.30, there is a significant $2^{nd}$ harmonic component in the drain current in Class-AB operation. On closer examination of the output side of the transistor, with a shorted output second harmonic, the drain voltage at $2f_0$ becomes zero based on a voltage-controlled drain-current source leading to a zero output power at the second harmonic. As a result, the drain voltage at the fundamental frequency becomes enlarged with steeper slopes and appears rather sinusoidal compared with the case without output second HT according to Figure 6.38(a). Consequently, the increased output power at $f_0$ due to the forced zero output power at the second harmonic leads to an improvement in efficiency. Another effect of the output second harmonic termination is that the overlap between the output current and output voltage becomes slightly reduced due to a smaller effective operating current angle $\alpha$ as can be seen in the $I_{DS}$ characteristic. This overlap indicates the power that is dissipated to heat by the device [6]. As a consequence, the less power that is dissipated by the device, the higher the PAE of the device becomes. Therefore, the PAE enhancement of the device with optimized $\Gamma_{L,PAE}(f_0)$ and output second harmonic tuning can also be explained by the reduced overlap between the intrinsic drain current and the intrinsic drain voltage.

The same conclusion can be drawn in case of a device with optimized $\Gamma_{L,PAE}(f_0)$ and input second harmonic termination, see Figure 6.38(b). In contrast to the output HT, input harmonic tuning has no significant influence on the drain voltage, being only indirectly influenced due to the modification of the drain current from the gate side. Consequently, there is almost no change in the $V_{DS,\,int}$ signal with and without input $2^{nd}$ HT. However, the modified drain current leads to a small effect on the PAE. The drain current characteristic in the figure shows a reduced operating angle $\alpha$ where $I_{DS} \geq 0$. Therefore, the overlap between the output current and output voltage becomes reduced, leading to less dissipated power and consequently to an improved PAE performance.

These results confirm the conclusion of [22] which indicates that the drain efficiency can be maximized by satisfying one of the two following equivalent conditions: maximize the fundamental output power or minimize both dissipated power on the device and the output harmonic power by appropriate selection of harmonic terminations. While the output HT is designed to properly terminate the drain current harmonic components with the aim of maximizing the fundamental output power, the input HT network controls the drain current generating mechanism as explained in [21].

Figure 6.39 shows the simulated dynamic intrinsic load curves for the device terminated with an optimized load at the fundamental and with the second harmonic set to 50 Ω, and optimized at the input and output, respectively. Again, all simulations are performed at 10 GHz, $P_{in}$= 32 dBm, and a biasing of $V_{DS}$= 30 V and $V_{GS}$= -2.25 V. In particular the loadline with output harmonic tuning shows a long vertical portion near the knee voltage corresponding to the relatively flat bottom of the drain voltage waveform (Figure 6.38(a)). In addition, the trajectory of the device without harmonic tuning is located far from the DC bias position ($V_{DS}$= 30 V and $I_{DS}$= 100 mA), while the trajectories of the device with optimized input or output harmonic tuning are kept close to or even around the location of DC bias. Input and output harmonic termination result in a modified load-line towards a Class-AB push-pull operation where the input and output harmonic termination emulates the $2f_0$ load of the second FET in contrast to the simulated loadline of the device without harmonic termination that

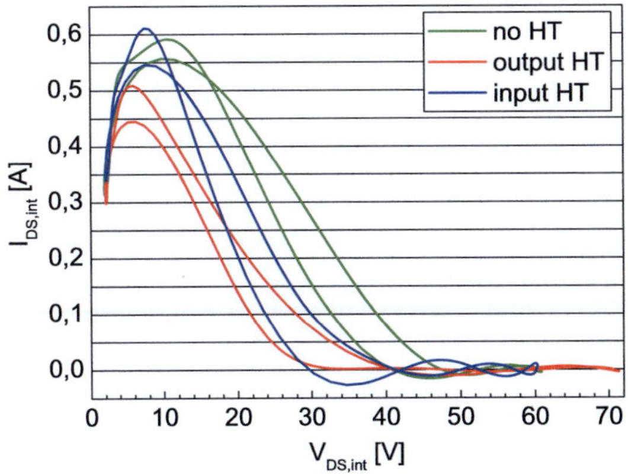


Figure 6.39: Simulated dynamic load lines at the intrinsic device level with optimized load at the fundamental. Second harmonic loads are either set to 50 Ω (no HT) or optimal tuned at the input or output of the device. All simulations are performed at 10 GHz, $P_{in}$= 32 dBm, and a biasing of $V_{DS}$= 30 V and $V_{GS}$= -2.25 V.

shows a typical Class-AB single-ended operation. The theoretically achievable PAE with a push-pull Class-B operation yields 78.5 % as shown in Chapter 3.1.2. However, this high PAE value is not possible in this case because of the knee voltage, Class-AB operation, and only $2^{nd}$ harmonic tuning instead of tuning all even harmonics.

### 6.2.2 Harmonic Short Circuit Stubs

The goal of a $2^{nd}$ harmonic short-circuit stub is the termination of the second harmonic while keeping optimal conditions for the fundamental matching, i.e. a high Q-factor at $f_0$ according to [112]. Different kinds of short-circuit structures exist. The most simple realization of a $2^{nd}$ harmonic short-circuit stub is an open-ended $\lambda/4$-transmission line. The drawback of this type is the fact that this structure offers no broadband application and additionally occupies a large chip area at X-band frequencies which makes a realization in MMIC technology difficult. For a wavelength of

$$\lambda(f_0) = \frac{c_0}{f_0 \cdot \sqrt{\varepsilon_{\text{r,eff}}}} \approx 11.8\,mm \tag{6.21}$$

at the fundamental frequency of $f_0$=10 GHz and with $\varepsilon_{\text{r,eff}}=6.5$ according to Section 5.3, the required quarter wavelength at $2f_0$ yields about 1.5 mm. The simulated input reflection coefficient of the $\lambda/4$-transmission line yields $Z_{11}(f_0) = 1.25\Omega - j86.65\Omega$ at 10 GHz leading to a Q-factor of $Q = \text{Im}(Z_{11}(f_0))/\,\text{Re}(Z_{11}(f_0)) = 69$. The parallel resistance results from the product of $\text{Im}(Z_{11}(f_0))$ and the Q-factor leading to 12.7 $k\Omega$.

However, the Q-factor strongly depends on the characterization of the losses in the MSL definitions and in case of LC-networks of the dielectric losses in the MIM capacitances. Since the characterization of the microstrip lines and MIM capacitances are not exactly defined, but only estimated, the value of the Q-factor is only estimated as well. Nevertheless, all simulated Q-factors in this section are calculated on the basis of identical assumptions. Therefore, these values only serve as a comparison between the different network variants.

To overcome the problem of the placement of such a large short-circuit stub-struc-ture, this section describes two different $2^{nd}$ harmonic short-circuit stub-structures that require less chip area, namely series resonators. Both short-circuit designs are analyzed in the following emphasizing the advantages and drawbacks of each configuration. The series resonators are then implemented in a shunt connection close to the input and/or output of the HEMT device according to the circuit block diagram of the single-stage HPA in Figure 6.31 to shorten the $2^{nd}$ harmonic at the gate and/or drain side of the device. Both short-circuit stubs are designed for best transmission at the fundamental frequency of 10 GHz and for best reflection at the $2^{nd}$ harmonic at 20 GHz.

### Stub-Structure 1

The first second-harmonic short-circuit stub consists of a relatively small MIM capacitor of 70 fF and a long short-circuit microstrip transmission line with $L_1$=0,9 nH. The resulting resonance frequency of this resonant circuit yields

$$f_{\text{res}} = \frac{1}{2\pi \cdot \sqrt{L_1 C_1}} \approx 20\,GHz. \tag{6.22}$$

The schematic and the layout of the stub-structure 1 are illustrated in Figure 6.40.

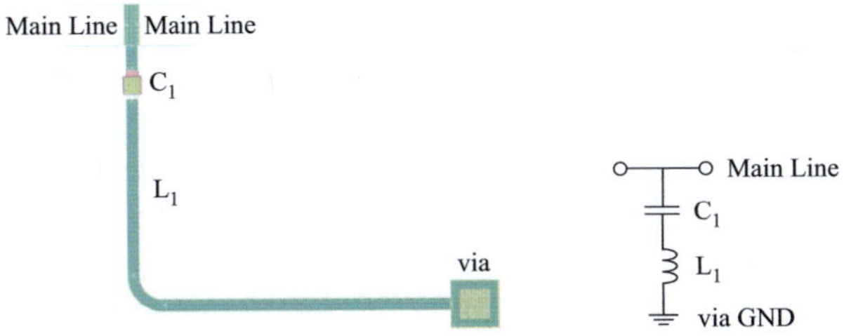


Figure 6.40: Schematic and layout of stub-structure 1 to shorten $2f_0$.

The simulated reflection coefficient $S_{11}$ of the stub-configuration 1 features only a small return loss of approximately -0.1 dB at 10 GHz and the desired resonance at 20 GHz with a return loss of better than -1.5 dB for a frequency range of about 4 GHz, as can be seen in Figure 6.41(a). In addition, $S_{11}$ in Figure 6.41(b) clearly reaches the required close proximity to the short region of the Smith chart at 20 GHz ($S_{11}(2f_0)$=0.81 $\angle$179.2°), but with some considerable losses and subsequently degradation of the expected power-added efficiency. According to Table 6.11, with this short-circuit stub-configuration with a $S_{11}$ magnitude of 0.81 at $2f_0$, a maximum PAE improvement of 3 % at device level can be realized, independent whether the $2^{nd}$ harmonic tuning is applied at the input or output of the HEMT device, respectively.

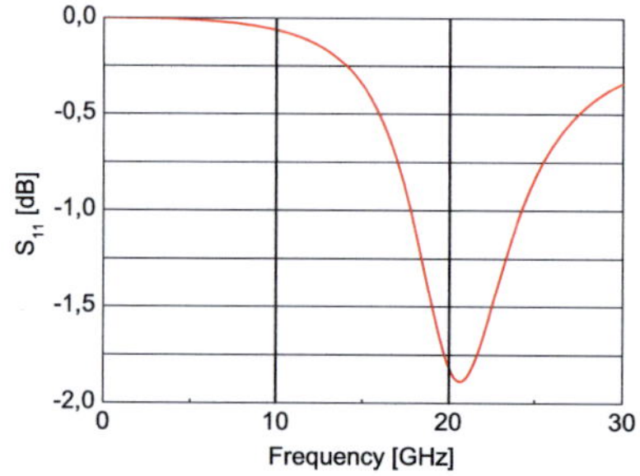


(a) $S_{11}$ in dB

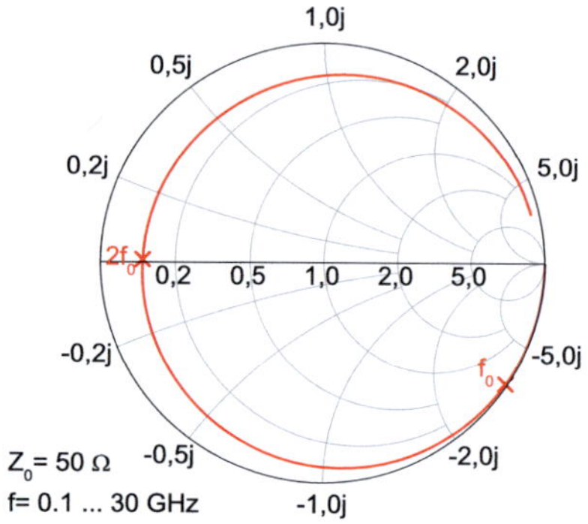


(b) $S_{11}$ in Smith chart

Figure 6.41: One-port simulation results of the short-circuit stub-structure 1.

Ideally, the additional short-circuit stub-network does not interact with the fundamental matching circuit, but this is only true if $S_{11}$ delivers an open at $f_0$=10 GHz with $1\angle 0°$. Nevertheless, it is not possible to design a 2-element series resonator at $2f_0$ with an open at $f_0$. For example, this stub-configuration yields a $S_{11}(f_0)$ at the fundamental frequency of $0.99\angle 33.7°$ (figure 6.41(b)). Therefore in practice a significant compensation readjustment to the fundamental matching network is necessary.

Furthermore the stub is optimized for a high Q-factor at the design frequency. The Q-factor is 76.7 for $Z_{11}(f_0) = 2.15\Omega - j165\Omega$ at 10 GHz. The resulting parallel resistance amounts $12.7\,k\Omega$. As a result, the Q-factor of this stub-structure is approximately 10 % higher than that of the $\lambda/4$-transmission line.

### Stub Structure 2

To reduce the significant loss of the stub-structure 1, a second short-circuit is designed. This stub-structure features two resonant circuits leading to two different resonance frequencies. The stub-circuit is designed for lower losses and to deliver "shorts" at the second and fourth harmonic, at 20 GHz and 40 GHz, respectively.

Figure 6.42 shows the schematic diagram and the layout of the stub-configuration 2. In this case, the lower short circuit with the larger dimensions of the MIM capacitance $C_1$ and the microstrip transmission line $L_1$ delivers the short at the second harmonic, while the upper short circuit with the smaller $C$ and $L$ dimensions causes the short at the $4^{th}$ harmonic. Both short-circuits are independent of each other and can be designed separately.

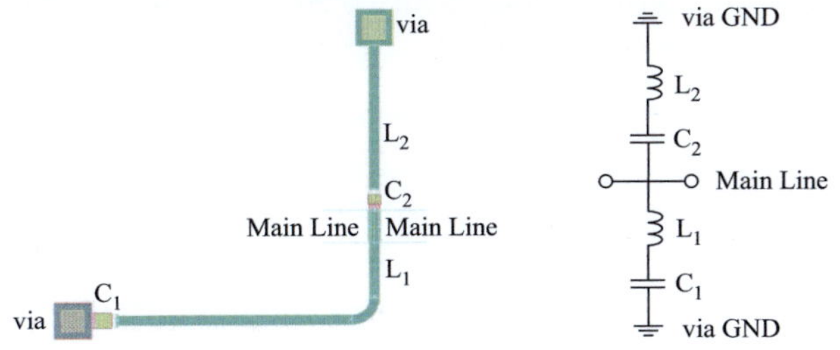


Figure 6.42: Schematic and layout of the stub-structure 2 to shorten $2f_0$ and $4f_0$.

The simulated reflection coefficient $S_{11}$ of the stub-structure 2 is depicted in Figure 6.43. This stub-structure features a similar reflec-

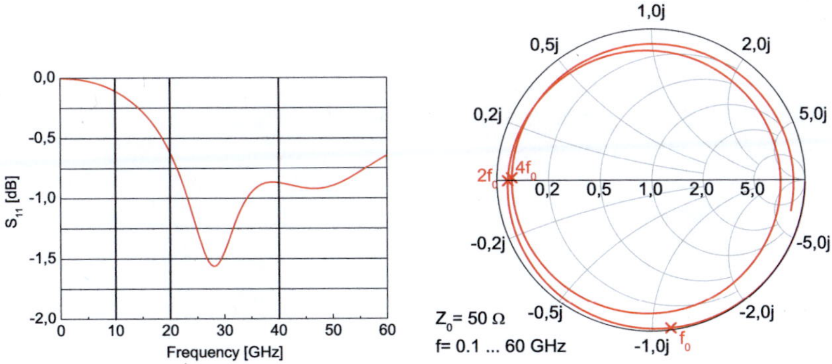


(a) $S_{11}$ in dB (b) $S_{11}$ in Smith chart

Figure 6.43: Simulated reflection coefficient $S_{11}$ of the short-circuit stub-structure 2.

tion at 10 GHz of approximately -0.1 dB compared with stub-configuration 1, while the reflection at 20 GHz differs a lot. On the one hand, at the second harmonic the short-circuit stub 2 delivers a three times lower reflection of approximately -0.5 dB. On the other hand, within the Smith chart the reflection coefficient $S_{11}$ clearly presents the required close proximity to the short region of the Smith chart at 20 GHz as well as at 40 GHz (see Figure 6.43(b)). The $S_{11}$ values for the fundamental frequency of 10 GHz as well as for the second and fourth harmonic are summarized in Table 6.12. The magnitudes for the $2^{nd}$ and $4^{th}$ harmonic are 0.93 and 0.90, respectively. These values correspond to a PAE enhancement of 5-5.5 % on device level as shown in Table 6.11.

The Q-factor of the stub-configuration 2 has about the same value at the fundamental frequency than the other stub-circuit, namely 75.5. However, the parallel resistance of the stub-circuit 2 amounts to 4.3 $k\Omega$. Thus, the resulting parallel resistance of the stub-structure 2 is about 3 times lower than that of the stub-configuration 1.

To sum up, the stub-configuration 1 yields a higher loss at the second harmonic leading to a smaller possible PAE improvement as shown in Table 6.11. Nevertheless, the resulting parallel resistance of this stub-circuit is three times higher than that of the stub-configuration 2. Consequently, stub-structure 1 implemented at the output of the device is barely discernible at $R_L$. In addition, stub-structure 1 requires less chip area and is therefore easier to implement at the input and/or output of the device.

Table 6.12: Simulated reflection coefficient $S_{11}$ of the entire stub-structure 2 at the fundamental frequency $f_0 = 10$ GHz and for the second and fourth harmonics.

| $S_{11}$ | **magnitude/phase** | **real+imaginary part of $\Gamma_L$** |
|---|---|---|
| at $f_0$ | $0.987 \angle -82.9°$ | $0.75\Omega - j56.6\Omega$ |
| at $2f_0$ | $0.93 \angle -179.8°$ | $1.79\Omega - j0.1\Omega$ |
| at $4f_0$ | $0.90 \angle +179°$ | $2.50\Omega + j0.45\Omega$ |

### 6.2.3 Effects of Harmonic Load Termination

The harmonic termination stub-structures should be implemented directly to the input and/or output of the active device for best results. Since this is not possible for the layout of the power amplifier, the closest accessible point outside of the transistor within the fundamental matching networks is used. This makes the harmonic short difficult to realize at the correct point resulting in the need for a readjustment of the stubs and fundamental matching networks.

Initially, the influence of an output harmonic termination on the fundamental output matching network is discussed concerning the two different stub-structures. Subsequently, the effect of an input harmonic termination on the fundamental input matching network is analyzed.

Harmonic load termination is a promising technique for realizing high efficiencies with MMIC power amplifiers. There are various publications that predict a PAE improvement of up to 10 % using harmonic load termination in power amplifiers within L-Band and X-Band frequencies in different technologies and operation classes [52], [4], [19], [36]. Terminating the second harmonic at the load of a transistor while keeping $\Gamma_{L,PAE}(f_0)$ at its optimum value leads to a PAE enhancement that results from the shaping of the drain current waveform [54].

The harmonic load termination in terms of a series resonant circuit tuned to the second harmonic frequency is shown schematically in a simplified active device output model without DC supply networks in Figure 6.44.

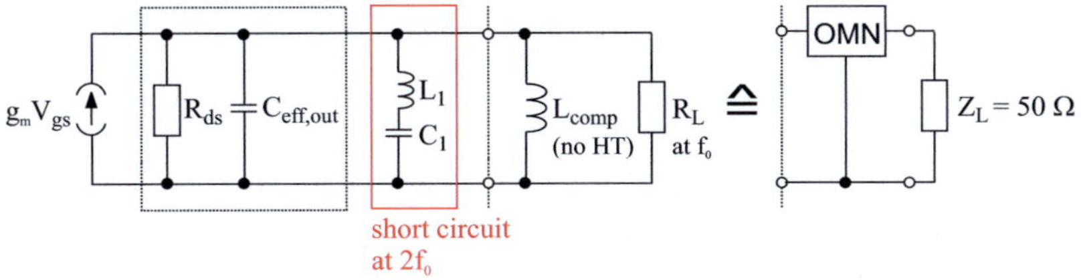


Figure 6.44: Simplified active device output model with second harmonic tuning.

The equivalent output circuit of the GaN HEMT can be approximated with the parallel-circuit of the drain-source resistance $R_{ds}$ and $C_{eff,out}$ with

$$C_{eff,out} = C_{ds} + C_{gd} \approx 0.4\,\mathrm{pF/mm} \tag{6.23}$$

as a typical value for a 0.25 $\mu$m GaN HEMT technology as considered here. Here $R_{ds}$ and the drain-source capacitance $C_{ds}$ are both nonlinear functions of the drain-source and gate-source bias voltages $V_{ds}$ and $V_{gs}$. Since $R_{ds} \gg R_L$ in the $I_{DS}$ saturation region it can be neglected. Hence, the load resistance $R_L$ can be calculated according to Equation 3.13 and yields about 60-100 Ω with an expected maximum drain current of $I_{max}$=0.8-1 A and a drain voltage of $V_{DS}$=30-40 V. Furthermore, at the fundamental frequency of 10 GHz, the reactance of $C_{eff,out}$ yields

$$X_C(f_0) = \left|\frac{1}{j\omega_0 C_{eff,out}}\right| = \frac{1}{2\pi f_0 C_{eff,out}} \approx 40\,\Omega\,, \tag{6.24}$$

i.e. $R_L$ and $X_C$ are at a comparable impedance level.

In addition, Figure 6.44 includes the short-circuit at $2f_0$ as well as the effective compensation inductance $L_{comp,\,no\,HT}$ connected in parallel to the load resistance $R_L$ at the fundamental. $L_{comp,\,no\,HT}$ parallel connected to $R_L$ is equivalent to the circuit on the right, consisting of the fundamental output matching network and the 50 Ω load.

The main problem of harmonic load termination of an AlGaN/GaN HEMT is the small value of $C_{eff,out}$ leading to a rather high impedance level of $\Gamma_L$ which makes a matching of the second harmonic short-stub at the load of the transistor more difficult than at the input of the device with the low input impedance $\Gamma_{in}$ of the HEMT. Thereby, the effective compensation inductance of the output matching network at the fundamental results in

$$L_{comp,\,no\,HT} = \left|\frac{X_C}{j\omega_0}\right| \approx 637\,pH\,. \tag{6.25}$$

The second harmonic short-circuit is realized through a series resonance of $C_1$ and $L_1$ to prevent the harmonic drain voltage component at $2f_0$ leading to a zero output power at the second harmonic as explained in more detail in Section 6.2.2. Accordingly, the drain voltage at the fundamental frequency becomes enlarged resulting in a higher output power at $f_0$ which in turn leads to a PAE enhancement.

The primary requirement on the short-circuit is a low-resistance short at $2f_0$ according to the results from the LP-simulations of the HEMT device with $2^{nd}$ harmonic tuning (see Table 6.11). Here, the PAE improvement is strongly influenced by $\Gamma_{L,2f0}$. Therefore, the demand that $|\Gamma_{L,2f0}|$ tends towards 1 leads to a maximum possible PAE.

To achieve this requirement, the parallel resistance $R_{res,2f0} = Z_i/Q$ of the stub-circuit needs to be minimized. With a constant given Q-factor in MMIC technology, the parallel resistance becomes reduced for a lowered image impedance $Z_i = \sqrt{L_1/C_1}$ of the stub-circuit which is the case for a low inductance $L_1$ and a high capacitance $C_1$.

Nevertheless, according to Equations 6.26 and 6.27 a high capacitance and low inductance in the $2f_0$ resonator lead to additional capacitance $C_{add}$ at $f_0$ that needs to be compensated with the fundamental OMN.

$$X_{Cadd}(f_0) = \left| j\omega_0 L_1 - \frac{1}{j\omega_0 C_1} \right| \tag{6.26}$$

$$C_{add}(f_0) = \left| \frac{1}{j\omega_0 X_{Cadd}(f_0)} \right| \tag{6.27}$$

Thus, the effective output capacitance $C_{eff,\,out\,HT}$ at the fundamental including $C_{add}$ of the second harmonic short-stub results in

$$C_{eff,\,out\,HT}(f_0) = C_{eff,out} + C_{add} \tag{6.28}$$

leading to a reduced compensation inductance $L_{comp,HT}$ according to Equation 6.25. As a result, the higher the effective output capacitance including $C_{add}$, the smaller is the resulting parallel resistance of the short-circuit at the fundamental frequency and in turn the narrower becomes the possible bandwidth of the fundamental output matching network.

$$R_{res} = \sqrt{\frac{L_{comp,HT}}{C_{eff,\,out\,HT}}} \Big/ Q \tag{6.29}$$

In other words, the higher the resulting parallel resistance of the short-circuit at the fundamental frequency, the less it is discernible compared with $R_L$. According to the results of both stub-configurations, the short-stub 2 offers a three times higher parallel resistance based on the calculations given in Subsection 6.2.2 which makes this stub-network more suitable for the use as a harmonic load termination.

Nevertheless, in simulations both stub-structures are implemented to the output of the active device followed by the fundamental output matching network. In this way, the influence of both harmonic load terminations on the fundamental output matching as well as on the net insertion loss are examined and compared with each other.

### Impact on Fundamental Output Matching Network

Figure 6.45 depicts the simulated output reflection coefficient (left side) and the net insertion loss (right side) of the output matching network without (blue line) and with $2^{nd}$ harmonic termination stubs, i.e. with the implemented short-stub 1 (green curve) and short-stub 2 (red curve). Thereby, the net insertion loss of the OMN is calculated according to Equation 3.17. The green area in the left figure indicates the -10 dB matching-bandwidth of the output matching network without termination stub-network yielding a bandwidth of approximately 4.5 GHz. Even though both short-circuit stubs are designed for terminating the second harmonic only with best transmission at the fundamental frequency, there is a strong influence of the termination stubs on the -10 dB matching-bandwidth of the matching network leading to a decrease of the -10 dB matching-bandwidth of about 1/3. In ad-

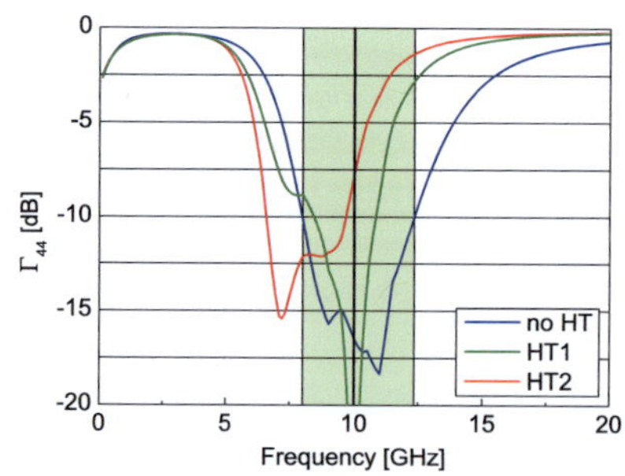


(a) Output reflection coefficient

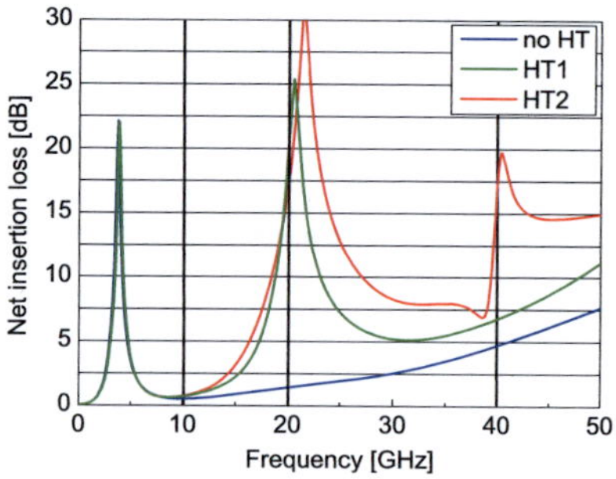


(b) OMN net insertion loss

Figure 6.45: Simulated output reflection coefficient $\Gamma_{44}$ and net insertion loss of the output matching network without (blue line) and with (red+green line) $2^{nd}$ harmonic terminations.

dition, the stub-structure 2 causes a frequency shift of the matching-bandwidth of about 2 GHz downwards that is only compensable with additional losses in bandwidth. Consequently, the implementation of a short-circuit stub at the output of an AlGaN/GaN HEMT device strongly affects the fundamental matching.

However, the matching networks with harmonic termination stub-circuits deliver a very high net insertion loss of more than 20 dB at the $2^{nd}$ harmonic (Figure 6.45(b)). The net insertion loss of the matching network without HT is less than 0.5 dB at the fundamental frequency, and increases only up to 0.65 dB taking one of the short circuit stub-structures into account. There is a good agreement between the increased net insertion loss and the reflection coefficient of the stub structure itself at 10 GHz (Figure 6.41(a)).

Due to the strong influence on the fundamental matching especially concerning the stub-circuit 2, the stub-structure 1 is chosen for the harmonic load termination. Even though the stub-configuration 1 yields the higher loss at the second harmonic leading to a theoretically smaller possible PAE improvement as shown in Table 6.11, the influence on the fundamental matching is less with regard to the high parallel resistance of this stub-circuit (see last subsection).

A comparison of the simulated reflection coefficients $\Gamma_{44}$ of the output matching network with (right side) and without (left side) second harmonic termination in terms of the stub-structure 1 is depicted in Figure 6.46. The green curve shows the output reflection coefficient $\Gamma_{44}$, while the red curve depicts the matching $S_{33}$ to the simulated optimum load impedance $\Gamma_{L,PAE}$ for maximum PAE of the power cell over a frequency range from 0.1 up to 30 GHz. The location for the optimum load reflection coefficient $\Gamma_{L,PAE}$ of the device at the fundamental frequency is displayed in blue. In case of an implemented $2^{nd}$ harmonic termination stub (Figure 6.46(b)), the $2^{nd}$ harmonic is closely matched to zero impedance in the Smith chart. $S_{33}(2f_0)$ yields $0.8\angle 178.8°$ compared to the stub-structure itself that delivers a magnitude of 0.81 at the second harmonic. The loss of the entire output matching network including the $2^{nd}$ harmonic stub-circuit is comparable to the loss of the stub-circuit at $2f_0$ itself. As shown in Table 6.11, theoretically a PAE improvement of 3.3 % is possible with this configuration.

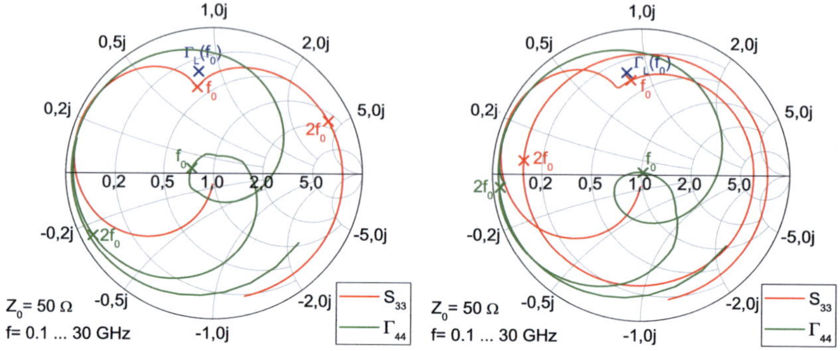


(a) no HT (b) with HT-1

Figure 6.46: Comparison of simulated reflection coefficients of the output matching network without (left side) and with (right side) second harmonic termination including the device's load reflection coefficient $\Gamma_L$ (blue) at fundamental frequency.

### 6.2.4 Influence of Input Harmonic Termination

In contrast to the output harmonic tuning, the input HT network controls the drain current generating mechanism as described in [21]. High efficiencies can be reached as long as the input signal driving the device current generator is purely sinusoidal [130]. However, in a real HEMT device with the transistor biased close to pinch-off voltage as in Class-AB and Class-B operation, the current waveform is non-ideal due to harmonic components leading to signal distortion. The main reason for the signal distortion at the input gate of the transistor is the typical gate capacitance $C_{gs}$ of the FET that yields a strong nonlinearity around threshold voltage $V_{th}$ with respect to applied input voltage. The non-linear characteristic of $C_{gs}$ distorts the input waveform thus destroying the a priori assumption of sinusoidal drive. The improvement of the performance limitation due to this effect is done by retrieving the sinusoidal drive by inserting short circuit input harmonic terminations directly in front of the device, ideally placed directly across $C_{gs}$ [137]. Initial simulations of the device with and without input HT are done to study the possible PAE improvement.

However, in the real HPA MMIC design it is not possible to place the input HT at its optimum position. PAE improvements due to the use of short-circuit stubs at the input of the transistor device are reported in [2], [65], [135], [38]. Most of these papers describe the achieved PAE values with input harmonic tuning based on Class-F operated pHEMT devices at frequencies between 1-5 GHz, however Watanabe [135] presents the influence of input HT on the basis of a GaAs FET performed at 6 GHz in Class-AB operation. A maximum simulated PAE of 31 % (no HT) and 38 % (with input HT) was achieved in saturation yielding to a PAE improvement of 7 %. This result has the same tendency to the here simulated AlGaN / GaN HEMT devices with and without input harmonic tuning given at the beginning of this section.

Alternative approaches to reduce the effect of non-linear input capacitance include a shunt capacitor circuit for realization of broader band solutions, but at the expense of worse input matching, due to the voltage-dependent increase of capacitance, proposed by P. White [138]. In addition, a novel compensation technique using an inverted diode is proposed by K. Jeon [49].

**Impact on Fundamental Input Matching Network**

In this section, the use of short-circuit input harmonic termination is analyzed in comparison to the short circuit output harmonic termination. Just as for the output matching network, both short-stub structures are inserted in front of the gate of the HEMT device during ADS simulations. Figure 6.47 shows the simulated input reflection coefficient (left side) and the net insertion loss (right side) of the input matching network without (blue line) and with $2^{nd}$ harmonic termination stubs, i.e. with the implemented short-stub 1 (green curve) and short-stub 2 (red curve). The green area in Figure 6.47(a) represents the -10 dB matching-bandwidth of the input matching network without harmonic termination covering a 3 GHz bandwidth. The comparison of the $\Gamma_{11}$-curves with and without HT shows only a small decrease in the -10 dB matching-bandwidth for the matching network with harmonic termination, whereas the sub-structure 2 (HT2) is insignificantly inferior to HT1. On the whole, the input reflection coefficient remains nearly unaffected by the termination stubs.

Regarding the net insertion loss of the input matching networks with and without $2^{nd}$ harmonic termination shown in Figure 6.47(b),

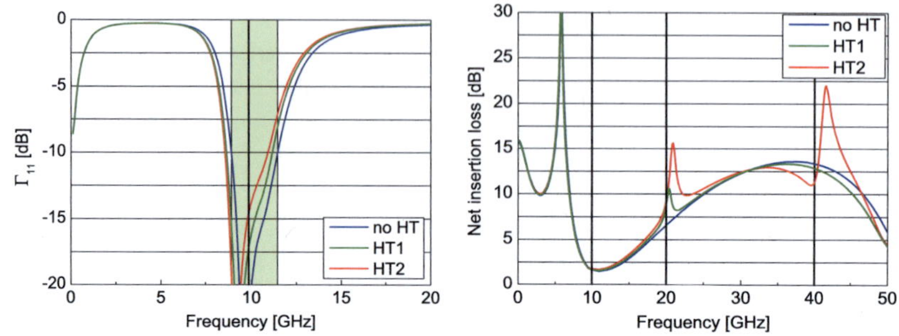


(a) Input reflection coefficient

(b) IMN net insertion loss

Figure 6.47: Simulated input reflection coefficient $\Gamma_{11}$ and net insertion loss of the input matching network without (blue line) and with (red+green line) $2^{nd}$ harmonic terminations.

as determined using Equation 3.19, the smallest net insertion loss is reached for a frequency range from the fundamental frequency of 10 GHz up to 12 GHz. At 10 GHz the net insertion loss yields 1.6 dB and 1.7 dB for the network without and with harmonic termination stub structures, respectively. This result fits perfectly together with the simulated reflection coefficient of the stub structure itself at the fundamental frequency shown in Figure 6.43(a). Nevertheless, a high net insertion loss at the $2^{nd}$ harmonic can be noticed for the network with harmonic termination stubs compared with the IMN without HT. The net insertion loss at the second harmonic becomes even more increased by the use of the short-circuit 2 that also features an additional high net insertion loss close to the $4^{th}$ harmonic.

In the ideal case, the input matching network is lossless, i.e. the net insertion loss is equal 0 dB. At the fundamental frequency of 10 GHz, the real network offers a minimum net insertion loss independent of the use of harmonic termination resonant short-stubs. However, it also shows high net insertion losses at $2f_0$ (HT1 and HT2) and $4f_0$ (HT2) due to the direct coupling of the "main line" and the resonant circuits.

Since both stub-circuits deliver similar results regarding the fundamental matching capabilities and the net insertion loss at 10 GHz, for further consideration the stub-circuit 2 is used for the input harmonic termination due to its lower loss at the second harmonic according to the last subsection.

Figure 6.48 gives a comparison of the simulated reflection coefficients of the input matching network with (right side) and without (left side) second harmonic termination using stub-structure 2. According to Figure 6.46, the green curve represents the matching of the input reflection coefficient $\Gamma_{in,opt}$ of the FET to the input load of 50 Ω at the fundamental frequency while the red curve displays the matching of the 50 Ω load at port 1 of the IMN to the conjugate $\Gamma_{in,opt}$ of the transistor. In addition, the location for $\Gamma_{in,opt}$ of the device at the fundamental frequency of 10 GHz is shown in blue.

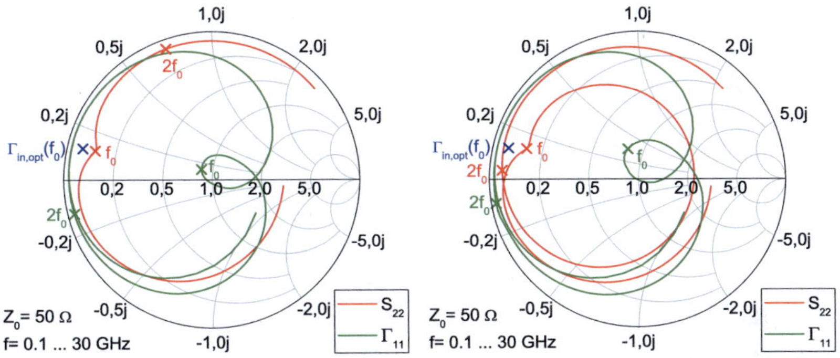


(a) without HT (b) with HT-2

Figure 6.48: Comparison of simulated reflection coefficients of the input matching network without (left side) and with (right side) second harmonic termination (stub-structure 2) including the device's opt. input reflection coefficient $\Gamma_{in,opt}$ (blue) at the fundamental frequency.

In Figure 6.48(a), the red curve shows a good matching to the optimum $\Gamma_{in,opt}$ at the fundamental, but no matching at the second harmonic ($S_{22}(2f_0) = 0.81\angle 166.2°$). However, in the case with an embedded termination stub (figure 6.48(b)), the red curve still shows a good matching to the optimum $\Gamma_{in,opt}$ at the fundamental, but also a very good shortening of the second ($S_{22}(2f_0) = 0.91\angle 179.8°$) and fourth ($S_{22}(4f_0) = 0.90\angle 179.3°$) harmonic. According to Table 6.12, the stub-circuit itself delivers a magnitude of 0.93 and 0.9 at $2f_0$ and $4f_0$, respectively. These values are comparable to the losses of the entire matching network at the second and fourth harmonic.

With this configuration, theoretically a PAE improvement of approximately 5.3 % is possible (see Table 6.11).

## Conclusions

Regarding the location of the opt. load reflection coefficient $\Gamma_{L,PAE}$ compared with the location for the optimum input reflection coefficient $\Gamma_{in,opt}$, a much higher design effort is necessary to reach an even worse matching result with smaller bandwidth for the harmonic termination at the output than at the input of the device. At 10 GHz, the high impedance of $\Gamma_{L,PAE}$ strongly affects the harmonic load termination compared with the low-impedance of $\Gamma_{in,opt}$ which is located close to the short in the Smith chart and offers only a small influence on the input harmonic termination. On the basis of the obtainable PAE values at device level, Table 6.13 gives an overview of the estimated PAE values for a single-stage PA terminated with $\Gamma_{L,PAE}$ at $V_{DS}$=30 V, $P_{in}$=32 dBm, and 10 GHz and in consideration of the influence of the above simulated fundamental matching networks including the input and output $2^{nd}$ harmonic terminations.

Table 6.13: Estimated PAE values for a single-stage PA terminated with $\Gamma_{L,PAE}$ at $V_{DS}$=30 V, $P_{in}$=32 dBm, and 10 GHz and in consideration of the above simulated fundamental matching networks including the input and output $2^{nd}$ harmonic terminations.

| **Fundamental IMN** | **Fundamental OMN** | **PAE** | **Δ PAE** |
|---|---|---|---|
| - | - | 54.2 % | - |
| - | with $2^{nd}$ HT | 57.5 % | +3.3 % |
| with $2^{nd}$ HT | - | 59.5 % | +5.3 % |
| with $2^{nd}$ HT | with $2^{nd}$ HT | 60.9 % | +6.7 % |

In addition, in contrast to the matching of the IMN to the optimum $\Gamma_{in,opt}$ the matching to the optimum $\Gamma_{L,PAE}$ in the output matching network is very sensitive towards even small changes of the $C_1$ and $L_1$ values within the short-circuit stub.

### 6.2.5 Design Configuration of the Monitor-HPAs

In this section, the use of $2^{nd}$ harmonic termination stubs at the input and/or output of a HEMT device is analyzed according to the applicability as a circuit design technique to improve the power-added efficiency of a high-power-amplifier. Since the PAE of a HPA mainly depends on the efficiency of the PA-stage as discussed in the last section (6.1), harmonic termination finds the highest impact when used at the input and/or output of the PA-stage transistors. Since in general the PA-stage of a X-band power-amplifier consists of several parallel FET cells, harmonic termination stubs need to be implemented symmetrically to all PA-stage FET cells. To identify the impact of the harmonic termination on the PAE performance of a HPA by excluding any other influencing factors, in this study only one FET cell is examined with an input and output matching network to a 50 $\Omega$ environment. Thereby, the PAE of the HEMT device together with the fundamental matching networks is analyzed with and without $2^{nd}$ harmonic termination at the input, output, and both input and output of the transistor [60].

#### Critical Design Aspects

As a result, four single-stage HPA MMICs, named Monitor 1 to Monitor 4, each with an 8x125 $\mu$m AlGaN/GaN-HEMT are designed with all possible $2^{nd}$ harmonic termination variations for a narrowband frequency range around the X-band center frequency of 10 GHz. For all MMICs the T-gate HEMT technology in chosen with a gate-to-gate pitch of 50 $\mu$m. No source-connected shields are implemented, and the aluminum concentration of the AlGaN layer is 22 % again.

To simplify measurements, the matching networks of these designs contain no additional DC-paths. The DC-biasing occurs beyond the RF-probes so that DC-probes are excluded leading to reduced complexity of the test structure MMICs. Nevertheless, the input and output matching networks contain parallel capacitances for DC-blocking. Though, all four power amplifiers are designed for comparison.

Table 6.14 gives an overview of the four PA designs with and without harmonic termination at the input and/or output of the device. The MMIC Monitor 1 consists only of the above mentioned HEMT device with fundamental matching networks at the input and output

Table 6.14: Overview of the $2^{nd}$ harmonic termination within the four PA MMICs.

| Circuit | HT within IMN | HT within OMN |
|---|---|---|
| Monitor 1 | - | - |
| Monitor 2 | - | HT |
| Monitor 3 | HT | - |
| Monitor 4 | HT | HT |

of the device but without any second harmonic terminations. Harmonic termination in form of additional short-circuit stubs is used within the output matching network (HPA Monitor 2), within the input matching network (HPA Monitor 3), and at the combination of both (HPA Monitor 4).

#### Layouts

The MMIC chip layouts of the HPA Monitor 1 to Monitor 4 are shown in Figure 6.49. The chip size of the MMIC Monitor 1 amounts to $2.25 \times 1.25\,\text{mm}^2$. With a chip size of $2.25 \times 1.5\,\text{mm}^2$ each, the other three designs are slightly larger than their pendant Monitor 1 that has no specific harmonic termination network. All four MMICs are processed on the same wafer for easy comparison.

### 6.2.6 Evaluation of the Harmonic Termination Influence on PAE

#### Small-Signal Measurements

Figure 6.50 shows a comparison of the measured CW-S-parameters for all four MMICs Monitor 1 to Monitor 4 at $V_{DS}$=30 V and a drain current $I_{DS}$=150 mA. The green region presents the 1 dB bandwidth of about 2 GHz relating to the original MMIC Monitor 1. As can be clearly seen in the figure, there is a reduction in the $S_{21}$-bandwidth for both MMICs with harmonic load termination (Monitor 2 and Monitor 4), while the MMIC with input harmonic termination (Monitor 3) retains the bandwidth of the MMIC Monitor 1 almost unchanged. The decrease in the $S_{21}$-bandwidth for the MMICs with output harmonic termination can be explained by the simulated fundamental matching

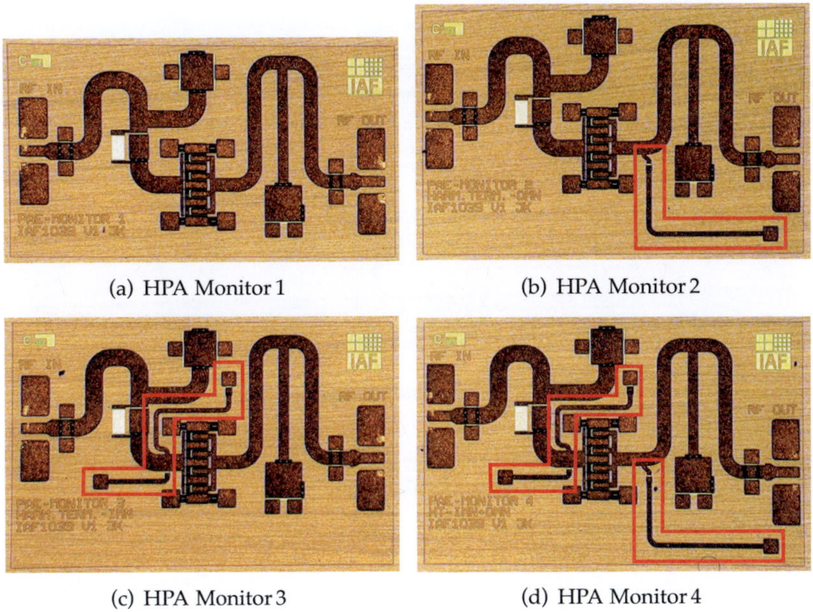

(a) HPA Monitor 1 (b) HPA Monitor 2

(c) HPA Monitor 3 (d) HPA Monitor 4

Figure 6.49: Photographs of the MMICs Monitor 1 to Monitor 4 with the harmonic termination networks marked. The chip size is $2.25 \times 1.25\,\text{mm}^2$ for the HPA Monitor 1 and $2.25 \times 1.5\,\text{mm}^2$ for the three other designs, respectively.

of the single networks with and without harmonic termination (see Subsection 6.2.3). There is a similar reduction in the -10 dB-matching-bandwidth of the corresponding output matching network with and without termination, while the -10 dB-matching-bandwidth of the input matching network with and without termination remains constant.

### Large-Signal Data

In Figure 6.51 the CW-power-results of all HPAs (Monitor 1-4) measured at $V_{DS}$=28 V and $V_{GS}$=-2.5 V are compared with each other. On the left side in Figure 6.51(a), CW-power sweeps at 10 GHz are shown, while on the right side in Figure 6.51(b) CW-power-measurements are performed for a frequency range from 8 GHz up to 11 GHz at a con-

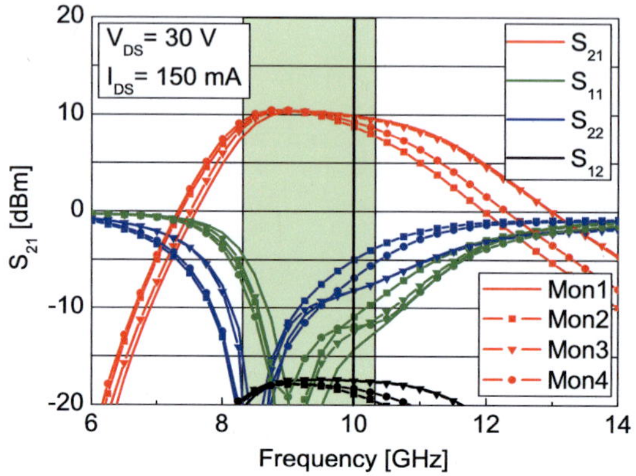


Figure 6.50: CW-S-parameter measurements of all HPAs Monitor 1-4. Bias point: $V_{DS}$=30 V and $I_{DS}$=150 mA.

stant input power of $P_{in}$=29 dBm. The input power level of 29 dBm is found to be the best to achieve maximum efficiency as can be seen in the figure on the left. The MMICs are measured in a broadband system with an input and output load of 50 Ω. As intended and designed, 10 GHz is found to be the optimum frequency to achieve maximum PAE with all four HPAs.

On the one hand, the achievable output power $P_{out}$ and gain at 10 GHz are reduced by up to 20 % and 9 % due to the harmonic termination as shown in Figure 6.51(a), respectively. Hereby, the output power $P_{out}$ is in the range from 36 dBm to 36.8 dBm for all HPA MMICs at an input power of $P_{in}$=29 dBm and at a frequency of 10 GHz. The associated gain reaches 7.2 - 7.8 dB for all HPAs.

On the other hand, the PAE is increased due to the application of harmonic termination despite of the reduced output power to the same time. The PAE enhancement implicates that the DC power must be reduced even stronger than the output power. For the HPA Monitor 1 without any intentional harmonic matching, a maximum PAE of 46.9 % is realized. Additional harmonic load termination (Monitor 2) and input harmonic termination (Monitor 3) give significant increases in the PAE up to 51.1 % and 52.2 %, respectively. The highest PAE of 53.2 % is measured with the MMIC Monitor 4 with both input and output harmonic termination. Furthermore, a strong reduction in

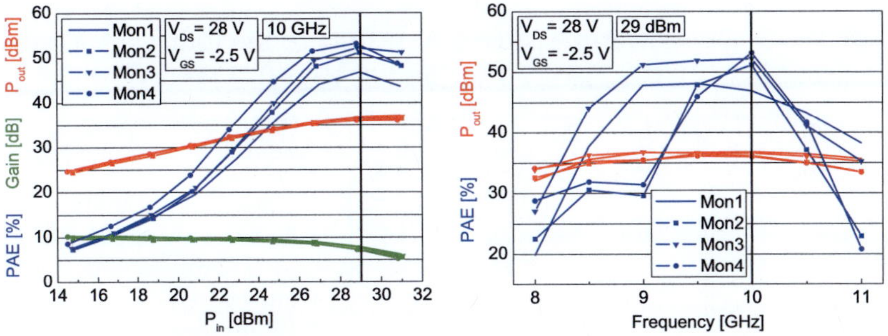


(a) Power sweeps at 10 GHz (b) Frequency sweeps at $P_{in} = 29$ dBm

Figure 6.51: Comparison of CW-power-measurements of all HPAs Monitor 1-4. Bias point: $V_{DS}$= 28 V and $V_{GS}$= -2.5 V.

PAE is shown in Figure 6.51(b) when moving away from the center frequency of 10 GHz. While Monitor 3 offers almost the same frequency range than the standard harmonically-unmatched Monitor 1, Monitor 2 and Monitor 4, both with output HT, show only an improvement of PAE for a small frequency range of less than 1 GHz from about 9.5 GHz up to nearly 10.5 GHz.

### Discussion

Both the power-measurements and the small-signal simulations and measurements show equivalent gain performances. While the input HT has almost no impact on the gain characteristic and matching over the entire X-band frequency range, the output HT strongly affects both the gain characteristic and the fundamental matching performance, especially towards higher X-band frequencies. The CW-power-measurement results are summarized in Table 6.15 for easy comparison.

The same as for the gain and matching performance accounts for the output power realized with the different HPA MMICs. On the one hand, input HT delivers only a small decrease in $P_{out}$ of 0.2 dB at 10 GHz compared with the standard HPA MMIC. On the other hand, output HT provides a reduction of 0.7 dB in the output power.

On closer examination of the obtainable PAE, the observed measured PAE improvements correspond perfectly to the estimated PAE

Table 6.15: CW-power-measurement results of the four HPA MMICs measured at 10 GHz, $P_{in}$=29 dBm, and at $V_{DS}$=28 V and $V_{GS}$=-2.5 V.

| MMIC | Gain | $P_{out}$ | PAE | Evaluation |
|---|---|---|---|---|
| Monitor 1 | 7.8 dB | 36.8 dBm ≡ 4.8 W | 46.9 % | - standard<br>- harmonically-unmatched |
| Monitor 2 (output HT) | 7.2 dB | 36.1 dBm ≡ 4.1 W | 51.1 % | - PAE-increase at 10 GHz > 4 % with reduced Gain and $P_{out}$<br>- PAE-increase over narrow bandwidth of 1 GHz<br>- high design effort |
| Monitor 3 (input HT) | 7.7 dB | 36.6 dBm ≡ 4.5 W | 52.2 % | - PAE-increase at 10 GHz > 5 % with const. Gain and slightly reduced $P_{out}$<br>- 4-5 % PAE-increase over 2.5 GHz bandwidth compared with standard |
| Monitor 4 (input+ output HT) | 7.2 dB | 36 dBm ≡ 4 W | 53.2 % | - PAE-increase at 10 GHz > 6 % with reduced Gain and $P_{out}$<br>- PAE-increase over narrow bandwidth of 1 GHz<br>- high design effort |

enhancements from Table 6.13. This study shows that the elaborated use of input as well as output second harmonic termination of an AlGaN/GaN HEMT delivers a comparable PAE improvement of 4-5 % on circuit level at the fundamental frequency of 10 GHz in CW-operation compared with the standard MMIC that is harmonically-unmatched. Using both input and output HT yields even a PAE enhancement of up to 6 %. This suggests the conclusion that with a typical 0.25 $\mu$m AlGaN/GaN HEMT technology, such as that analyzed here, a PAE-enhancement of up to 6 % over a narrow bandwidth is realistic as long as the technology is stable and there are not too many devices in parallel in the PA-stage.

Nevertheless, the design of a suitable network for broadband harmonic termination is a challenge. Within this section, the potential for HPAs with high PAE-values is evaluated only for narrow-band applications. As can be seen in Figure 6.51(b), the PAE improvement due to output second harmonic termination is realized only at the expense of smaller bandwidth. In this case, an increase in PAE can be obtained only over a narrow bandwidth of 1 GHz around the fundamental frequency, i.e. a relative bandwidth of 10 %. There is a strong decrease in the obtainable PAE below and above this frequency range. In contrast to the output HT, the input harmonic termination delivers a PAE enhancement of 4-5 % over a wider bandwidth of about 2.5 GHz around the fundamental frequency compared with the standard MMIC. As a result, using input harmonic termination delivers a high potential for further HPA applications even with the purpose of broader bandwidth of up to 2-3 GHz.

# Chapter 7

# Conclusion and Outlook

GaN based HPAs have become of particular importance for airborne and space related radar applications as well as for communication systems [95], [46]. Currently, however, the GaN-based transistor technology development is still in progress, offering lower power-added efficiency (PAE) values in comparison to e.g. GaAs [18]. Due to the limited power supplies of modern airborne and space related self-sustaining systems, PAE improvement is indispensable. As a result, the focus of this thesis was the design of AlGaN / GaN-HEMT high-power amplifiers in Class-AB operation for X-band (8-12 GHz) applications with particular emphasis on an optimized PAE. Class-AB operation is commonly used for high-power radar applications where high output-power levels above 20 W with associated PAE values beyond 40 % are required.

Figure 7.1 gives a temporal overview of the RF-performance of all single- and dual-stage HPA MMICs that were designed, processed, and evaluated in the course of this work. All HPAs were designed for a center frequency between 8 - 10 GHz with main attention to the realizable PAE performance. Besides the indicated values that were measured in CW-mode, all other MMICs were operated in pulsed-RF mode with long-pulse operation (100 $\mu$s) for a duty cycle of 10 %. Especially the single-stage HPAs (unfilled symbols) suggest a continuous development with significant PAE improvements on MMIC design level over time. Essential PAE enhancements could be obtained related to the first HPA designs that offered PAE values of about 20 - 25 %. In the course of this work, PAE values of above 40 % could be

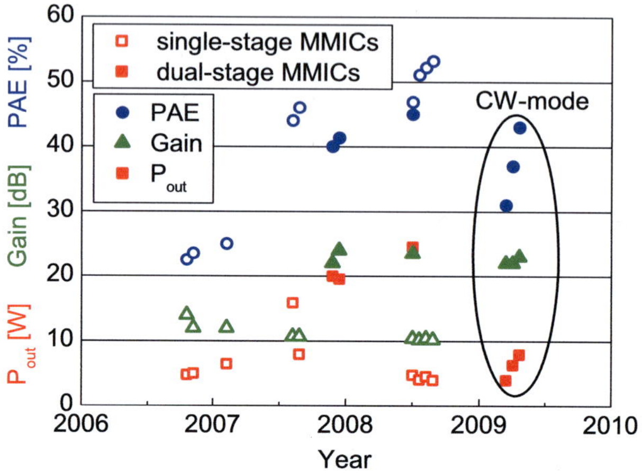


Figure 7.1: Temporal overview of the RF-performances of all HPA MMICs reported in this work with a center frequency between 8-10 GHz. Besides the indicated circuits that were measured in CW-mode, all other MMICs were operated in pulsed-RF mode with a duty cycle of 10 %.

realized with recent dual-stage HPA MMICs in pulsed-RF and even in CW-mode, and PAE values of even above 50 % were measured with single-stage monitor amplifiers.

This significant increase in PAE over time was achieved due to both a PAE enhancement at the device level and on improved design techniques. The first aspect consists mainly in the GaN HEMT device processing technology. Over the last few years, the DC- and RF-performance of the AlGaN/GaN HEMT could be increased rapidly up to the outstanding Fraunhofer IAF 0.25 $\mu$m GaN HEMT technology performance demonstrated in Chapter 2.1. The impact of the technology progress on HPA circuit designs was investigated within the scope of this work (Chapter 4). By means of modifications on the gate module, i.e. the implementation of gate- and source-connected field modulating plates, as well as due to an improved HEMT structure and layout, i.e. the reduction of the aluminum mole fraction and a reduced gate-to-gate pitch, the PAE could be improved from 22.5 % to over 40 % at the HPA design level.

The second aspect, the PAE improvement on the design level, was of particular interest in this thesis. In addition to the different design aspects for PAE enhancement, this work presented a first balanced microstrip HPA demonstrator in GaN technology for X-band frequencies to overcome the GaN-specific problem of a high load-impedance level that causes a poor output matching capability. So far, this is the only balanced AlGaN/GAN HEMT HPA available in microstrip transmission line technology. It was designed in a modular concept taking advantage of the GaN-specific high load-impedance level. As a result, with the balanced HPA design a strongly improved input and output matching capability of better than -14 dB could be obtained over the entire bandwidth from 8.5 - 13 GHz compared to -5 dB for the single-ended approach. Further optimizations of this balanced approach promise the great potential to further improve the RF-performance of the HPA and reduce the required MMIC chip size to the same time.

However, mainly two different design features for PAE improvement were analyzed in theory and proven on the basis of single- and dual-stage high-power amplifiers designed for Class-AB operation at X-band frequencies. These two design aspects were investigated by means of two highly sophisticated design studies.

The first study focused on a PAE improvement due to a skillful selection of the power (PA)-stage to driver (DRV)-stage gate-width ratio of dual-stage HPA MMICs. Most GaN-based power amplifier designs published in the literature use PA/DRV TGW ratios of about 2:1 [23], [102]. Nevertheless, a fundamental theoretical analysis indicated that higher PA/DRV TGW ratios of up to 4:1 offer promising performance results concerning the PAE behavior of these HPAs. This is due to the fact that a reduced total gate-width (TGW) of the DRV-stage at a constant PA-stage size requires less DC power. To achieve high efficiencies, the TGW of the DRV-stage needs to be designed as small as possible in order not to waste a lot of DC power which is not converted into output power. However, the DRV-stage still has to be sufficiently large to drive the PA-stage transistors into saturation even over the required bandwidth of the HPA.

Therefore, HPA MMICs with different PA/DRV TGW ratios between 2:1 and 4:1 were designed, developed, and characterized. The evaluation of the experimental results proved the theoretical assumption. An increased PAE could be obtained for higher PA/DRV TGW ratios but at the expense of a smaller realizable frequency bandwidth. Depending on the required output power, PAE, and operating frequency bandwidth, an increased PA/DRV TGW ratio turns out to be an effective design measure to improve the PAE. It could be demonstrated that a PA/DRV TGW ratio of 3:1 is feasible for X-band GaN HEMT HPAs with an output power level higher than 20 W and PAE values above 40 % over a frequency bandwidth of about 1.6 GHz for approximately 8-9.6 GHz. Compared to comparable HPA designs with a PA/DRV TGW ratio of 2:1, which so far is widely used for GaAs and even for GaN designs, a PAE improvement of up to 5 % was possible for a very narrow frequency range. However, the design study exemplified that a 4:1 TGW ratio is overdone even for narrow X-band GaN HPA designs.

A reduced PA/DRV TGW ratio could also be beneficial for future switching amplifier applications in communication systems operated in Class D, E, and F.

In the second analysis, the influence of the second harmonic of a GaN HEMT, being the largest harmonic besides the fundamental in Class-AB to Class-B operation, was examined. For both GaAs and GaN technologies, input harmonic termination (HT) is favored according to the literature [49], [137]. However, no or only rudimental reasons for the preferable use of input HT are mentioned there.

Due to the fact that the basic mechanism of second harmonic termination at both the input and output of a GaN-based transistor was not satisfactorily treated in the literature, it was fundamentally analyzed in theory within this work on the basis of load-pull simulations of AlGaN/GaN HEMT models. The theoretically achieved results offer a large potential for PAE enhancement. By time-domain waveform analysis of the large-signal simulations of input and output HT, the principal mechanisms of input and output HT could be clarified: output HT leads to output voltage shaping and input HT results in an input voltage shaping that in turn forms the output current.

Subsequent to the theoretical analysis on FET device level, single-stage power amplifiers were designed to prove the expected PAE increase due to the use of input and/or output second harmonic termination. As expected, a PAE improvement of 4-5 % could be successfully demonstrated at the center frequency of 10 GHz for either input or output second harmonic termination and up to 6 % using both input and output harmonic termination. However, at the output side the beneficial PAE increase was obtained only at the expense of a strongly reduced frequency bandwidth down to only about 1 GHz bandwidth and an approximately 0.5 dB lower output power. In contrast, second harmonic termination at the input side delivers the desired and expected PAE increase over a frequency bandwidth of 2-3 GHz at X-band frequencies without considerable loss in output power. Therefore, input harmonic termination is most suitable for HPA applications where high PAE values are required.

The results obtained with harmonic termination show similar effects compared to switching amplifiers concerning the signal forming at the input and output of the HEMT device. Both, the HT and switching amplifiers seem to have the same impact on a HPA.

Summarizing, the results of this work have the potential to essentially contribute to further AlGaN/GaN HEMT HPA design developments for next generation communication systems as well as for airborne and space related applications where high overall efficiencies are of particular necessity.

# Appendix A

# Calculation of HEMT-based Small-Signal Model Parameters

The small-signal equivalent circuit of a HEMT is shown in Figure 2.1 in Chapter 2.1. The intrinsic circuit model consists of eight nonlinear components that are directly dependent on the process-oriented technological parameters. They accurately predict the bias-dependent S-parameters. The equivalent circuit model elements can be extracted by generating the frequency-dependent Y-parameters of the intrinsic device. The expressions of the capacitance components can be calculated in terms of the Y-parameters [110]:

$$C_{\mathrm{gd}}(\omega)\,[\mathrm{pF/mm}] = \frac{-\mathrm{Im}[Y_{12}(\omega)]}{\omega}\cdot 10^{12}/\mathrm{TGW(mm)} \qquad (A.1)$$

$$\begin{aligned} C_{\mathrm{gs}}(\omega)\,[\mathrm{pF/mm}] &= \frac{\mathrm{Im}[Y_{11}(\omega)] - \omega C_{\mathrm{gd}}(\omega)}{\omega}\cdot 10^{12}/\mathrm{TGW(mm)} \\ &= \frac{\mathrm{Im}[Y_{11}(\omega) + Y_{12}(\omega)]}{\omega}\cdot 10^{12}/\mathrm{TGW(mm)} \qquad (A.2) \end{aligned}$$

$$\begin{aligned} C_{\mathrm{ds}}(\omega)\,[\mathrm{pF/mm}] &= \frac{\mathrm{Im}[Y_{22}(\omega)] - \omega C_{\mathrm{gd}}(\omega)}{\omega}\cdot 10^{12}/\mathrm{TGW(mm)} \\ &= \frac{\mathrm{Im}[Y_{22}(\omega) + Y_{12}(\omega)]}{\omega}\cdot 10^{12}/\mathrm{TGW(mm)}\,. \qquad (A.3) \end{aligned}$$

Furthermore, the transconductance $g_{\mathrm{m}}(\omega)$ and output conductance $g_{\mathrm{ds}}(\omega)$ are defined as follows

$$g_{\mathrm{m}}(\omega)\,[\mathrm{mS/mm}] = \frac{\mathrm{Real}[Y_{21}(\omega) - Y_{12}(\omega)]}{|1 + j\omega R_{\mathrm{gs}}\cdot 10^{-3} C_{\mathrm{gs}}\cdot 10^{-9}|}\cdot 10^{3}/\mathrm{TGW(mm)} \qquad (A.4)$$

$$G_{\mathrm{ds}}(\omega)\,[\mathrm{mS/mm}] = \mathrm{Real}[Y_{22}(\omega) + Y_{12}(\omega)]\cdot 10^{3}/\mathrm{TGW(mm)}. \qquad (A.5)$$

The last three intrinsic parameters yield

$$R_{\mathrm{gd}}(\omega)\,[\Omega\mathrm{mm}] = \mathrm{Real}\left[-\frac{1}{Y_{12}(\omega)}\right]\cdot \mathrm{TGW(mm)} \qquad (A.6)$$

$$R_{\mathrm{gs}}(\omega)\,[\Omega\mathrm{mm}] = \mathrm{Real}\left[\frac{1}{Y_{11}(\omega) + Y_{12}(\omega)}\right]\cdot \mathrm{TGW(mm)} \qquad (A.7)$$

$$\tau(\omega)\,[psec] = \frac{1}{\omega}\frac{\mathrm{Im}(Y_{12}(\omega) - Y_{21}(\omega))}{\mathrm{Real}(Y_{21}(\omega) - Y_{12}(\omega))}\cdot 10^{12}\,. \qquad (A.8)$$

# Appendix B

# Pulsed-RF Power Measurement Setup

In general, the large-signal measurements of the high-power amplifiers presented in this work are done in pulsed-RF mode at an ambient temperature of $T_a$= 25 °C. The configuration of the used pulsed-RF power measurement setup is shown in Figure B.1. The system consists of a probe station including a thermo chuck on which the device under test (DUT) is placed for the measurements. Bias networks combine the RF, DC, and pulsed stimulus from the network analyzer and the pulse bias sources, and deliver the signals to the DUT. Furthermore, the IC-CAP software extracts and displays the measurement results.

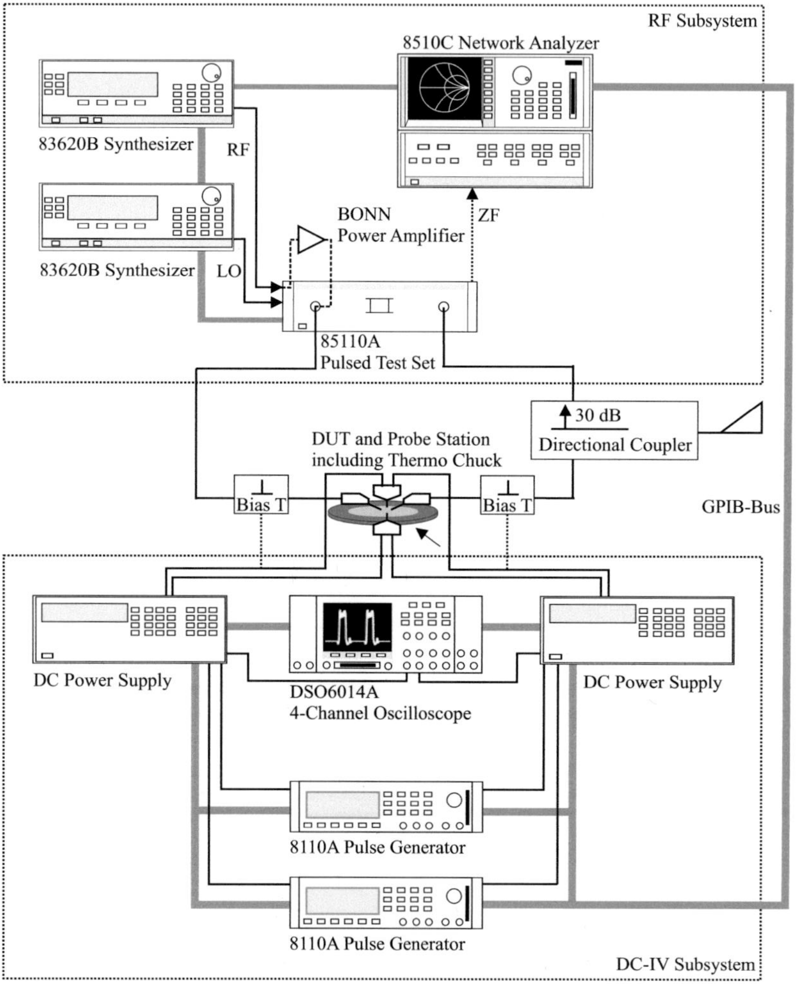


Figure B.1: Block diagram of pulsed-RF power measurement system.

# List of References


[1] P. Abrie, *The Design of Impedance Matching Networks for Radio-Frequency and Microwave Amplifiers.* Boston, London: Artech House, 1985.

[2] M. Akkul, M. Roberts, V. Walker, and W. Bosch, "High Efficiency Power Amplifier Input/Output Circuit Topologies for Base Station and WLAN Applications," in *IEEE MTT-S International Microwave Symposium Digest*, vol. 2, 2004, pp. 843 – 846.

[3] Y. Ando, Y. Okamoto, H. Miyamoto, T. Nakayama, T. Inoue, and M. Kuzuhara, "10-W/mm AlGaN-GaN HFET with a Field Modulating Plate," *IEEE Electron Device Letters*, vol. 24, no. 7, pp. 289 – 291, May 2003.

[4] H. Bae, R. Negra, S. Boumaiza, and F. Ghannouchi, "High-Efficiency GaN Class-E Power Amplifier with Compact Harmonic-Suppression Network," in *Proceedings of the 2nd European Microwave Integrated Circuits Conference, EuMC 2007*, Munich, Germany, Oct. 2007, pp. 1093 – 1096.

[5] M. Banu and V. Prodanov, "A System Approach to Integrated Power Amplifier Analysis and Design," in *ISSCS International Symposium on Signals, Circuits and Systems*, vol. 1, July 2007, pp. 1 – 4.

[6] D. Barataud, F. Blache, A. Mallet, P. Bouysse, J.-M. Nebus, J. Villotte, J. Obregon, J. Verspecht, and P. Auxemery, "Measurement and Control of Current/Voltage Waveforms of Microwave Transistors Using a Harmonic Load-Pull System for the Optimum Design of High Efficiency Power Amplifiers," *IEEE Transactions*

*on Instrumentation and Measurement*, vol. 48, no. 4, pp. 835 – 842, 1999.

[7] R. Behtash, H. Tobler, F. Berlec, V. Ziegler, H. Leier, B. Adelseck, T. Martin, R. Balmer, D. Pavlidis, R. Jansen, M. Neuburger, and H. Schumacher, "Coplanar AlGaN/GaN HEMT Power Amplifier MMIC at X-band," in *IEEE MTT-S International Microwave Symposium Digest*, vol. 3, 2004, pp. 1657 – 1659.

[8] A. Bessemoulin, R. Quay, S. Ramberger, H. Massler, and M. Schlechtweg, "A 4-Watt X-Band Compact Coplanar High-Power Amplifier MMIC With 18-dB Gain and 25 % PAE," *IEEE Journal of Solid-State Circuits*, vol. 38, no. 9, Sept. 2003.

[9] S. Binari, P. Klein, and T. Kazior, "Trapping Effects in GaN and SiC Microwave FETs," *IEEE Proceedings*, vol. 90, no. 6, pp. 1048 – 1059, 2002.

[10] D. Blackburn, "Temperature Measurements of Semiconductor Devices - A Review," in *IEEE Semiconductor Thermal Measurement and Management Symposium*, Sept. 2004, pp. 70 – 80.

[11] H. Bode, *Network Analysis and Feedback Amplifier Design*. New York: Van Nostrand, 1945.

[12] T. Boles, "Advanced Components for Applications in S-Band and X-Band Radars," in *Proceedings of the 3rd European Microwave Integrated Circuits Conference, EuMIC 2008*, Amsterdam, Netherlands, Oct. 2008, pp. 258 – 261.

[13] A. Brannick, N. Zakhleniuk, B. Ridley, J. Shealy, W. Schaff, and L. Eastman, "Influence of Field Plate on the Transient Operation of the AlGaN/GaN HEMT," *IEEE Electron Device Letters*, vol. 30, no. 5, pp. 436 – 438, May 2009.

[14] A. Chini, D. Buttari, R. Coffie, S. Heikman, S. Keller, and U. Mishra, "12 W/mm Power Density AlGaN/GaN HEMTs on Sapphire Substrate," *Electronics Letters*, vol. 40, no. 1, pp. 73 – 74, 2004.

[15] A. Chini, D. Buttari, R. Coffie, L. Shen, S. Heikman, A. Chakraborty, S. Keller, and U. Mishra, "High Performance AlGaN/GaN HEMTs with a Field Plated Gate Structure," in *IEEE*

International Semiconductor Device Research Symposium, 2003, pp. 434 – 435.

[16] C.-K. Chu, H.-K. Huang, H.-Z. Liu, C.-H. Lin, C.-H. Chang, C.-L. Wu, C.-S. Chang, and Y.-H. Wang, "An X-Band High-Power and High-PAE PHEMT MMIC Power Amplifier for Pulse and CW Operation," *IEEE Microwave and Wireless Components Letters*, vol. 18, no. 10, pp. 707 – 709, oct. 2008.

[17] R. Chu, L. Shen, N. Fichtenbaum, D. Brown, Z. Chen, S. Keller, S. DenBaars, and U. Mishra, "V-Gate GaN HEMTs for X-Band Power Applications," *IEEE Electron Device Letters*, vol. 29, no. 9, pp. 974 – 976, Sept. 2008.

[18] R. Chu, L. Shen, N. Fichtenbaum, Z. Chen, S. Keller, S. DenBaars, and U. Mishra, "Correlation Between DCŰRF Dispersion and Gate Leakage in Deeply Recessed GaN / AlGaN / GaN HEMTs," *IEEE Transactions on Electron Devices*, vol. 29, no. 4, pp. 303 – 305, 2008.

[19] Y. Chung, C. Hang, S. Cai, Y. Chen, W. Lee, C. Wen, K. Wang, and T. Itoh, "Effects of Output Harmonic Termination on PAE and Output Power of AlGaN / GaN HEMT Power Amplifier," *IEEE Microwave and Wireless Components Letters*, vol. 12, no. 11, pp. 421 – 423, 2002.

[20] R. Coffie, S. Heikman, D. Buttari, S. Keller, A. Chini, L. Shen, N. Zhang, A. Jimenez, D. Jena, and U. Mishra, "P-GaN / AlGaN / GaN High Electron Mobility Transistors," in *Device Research Conference Digest (DRC)*, 2002, pp. 25 – 26.

[21] P. Colantonio, F. Giannini, and G. L. et al., "Multiharmonic Manipulation for Highly Efficient Microwave Power Amplifiers," *International Journal of RF and Microwave Computer-aided Engeneering*, vol. 11, no. 6, pp. 366 – 384, 2001.

[22] P. Colantonio, F. Giannini, G. Leuzzi, and E. Limiti, "Harmonic Tuned PAs Design Criteria," in *IEEE MTT-S Proceedings of the International Microwave Symposium Digest*, vol. 3, Seattle, Wa, USA, June 2002, pp. 1639 – 1642.

[23] C. Costrini, M. Calori, A. Cetronio, C. Lanzieri, S. Lavanga, M. Peroni, E. Limiti, A. Serino, G. Ghione, and G. Melone, "A

20 Watt Microstrip X-Band AlGaN/GaN HPA MMIC for Advanced Radar Applications," in *Proceedings of the 3rd European Microwave Integrated Circuits Conference, EuMIC 2008*, Amsterdam, Netherlands, Oct. 2008, pp. 566 – 569.

[24] S. Cripps, "A Theory for the Prediction of GaAs FET Load-pull Power Contours," in *IEEE MTT-S International Microwave Symposium Digest*, May 1983, pp. 221 – 223.

[25] ——, *RF Power Amplifiers for Wireless Communications.* Boston, London: Artech House, 2006.

[26] M. Dammann, M. Cäsar, H. Konstanzer, P. Waltereit, R. Quay, W. Bronner, R. Kiefer, S. Müller, M. Mikulla, P. van der Wel, T. Rödle, F. Bourgeois, and K. Riepe, "Reliability Status of GaN Transistors and MMICs in Europe," in *IEEE International Reliability Physics Symposium (IRPS)*, 2010, pp. 129 – 133.

[27] M. Dammann, M. Cäsar, P. Waltereit, W. Bronner, H. Konstanzer, R. Quay, S. Müller, M. Mikulla, O. Ambacher, P. van der Wel, T. Rödle, R. Behtash, F. Bourgeois, and K. Riepe, "Reliability of AlGaN/GaN HEMTs under DC- and RF-operation," in *Reliability of Compound Semiconductors Digest (ROCS)*, 2009, pp. 19 – 32.

[28] A. Darwish, B. Huebschman, E. Viveiros, and H. Hung, "Dependence of GaN HEMT Millimeter-Wave Performance on Temperature," *IEEE Transactions on Microwave Theory and Techniques*, vol. 57, no. 12/2, pp. 3205 – 3211, 2009.

[29] A. de Hek, "Design, Realisation and Test of GaAs-based Monolithic Integrated X-band High-Power Amplifiers," Ph.D. dissertation, Dordrecht Polytechnic, Netherlands, July 2001.

[30] A. de Hek, G. van der Bent, M. van Wanum, and F. van Vliet, "A Cost-Effective 10 Watt X-band High Power Amplifier and 1 Watt Driver Amplifier Chip-Set," in *European Gallium Arsenide and Other Semiconductor Application Symposium*, Oct. 2005, pp. 37 – 40.

[31] M. Edwards and J. Sinsky, "A New Criterion for Linear 2-Port Stability Using a Single Geometrically Derived Parameter,"

*IEEE Transactions on Microwave Theory and Techniques*, vol. 40, no. 12, pp. 2303 – 2311, Dec. 1992.

[32] K. Eisele, R. Engelbrecht, and K. Kurokawa, "Balanced Transistor Amplifiers for Precise Wideband Microwave Applications," in *IEEE International Solid-State Circuits Conference*, Feb. 1965, pp. 18 – 19.

[33] F. Ellinger, *Radio Frequency Integrated Circuits and Technologies*. Berlin, Heidelberg, New York: Springer Verlag, 2007.

[34] R. Fano, "Theoretical Limitations on the Broadband Matching of Arbitrary Impedances," *Journal of the Franklin Institute*, vol. 249, no. 1, pp. 57 – 83, 1950.

[35] S. Feng, P. Hu, G. Zhang, C. Guo, X. Xie, and T. Chen, "Determination of Channel Temperature of AlGaN/GaN HEMT by Electrical Method," in *Semiconductor Thermal Measurement and Management Symposium, 2010. SEMI-THERM 2010. 26th Annual IEEE*, 2010, pp. 165 –169.

[36] D. FitzPatrick, J. Lees, A. Sheikh, J. Benedikt, and P. Tasker, "Systematic investigation of the impact of harmonic termination in the efficiency performance of above octave bandwidth microwave amplifiers," in *Proceedings of the 4th European Microwave Integrated Circuits Conference, EuMC 2009*, Rome, Italy, Oct. 2009, pp. 1445 – 1448.

[37] S. Gao, "High Efficiency Class-F RF/Microwave Power Amplifiers," *IEEE Microwave Magazine*, vol. 7, no. 1, pp. 40 – 48, 2006.

[38] S. Gao, P. Butterworth, S. Ooi, and A. Sambell, "High-efficiency power amplifier design including input harmonic termination," *IEEE Microwave and Wireless Components Letters*, vol. 16, no. 2, pp. 81 – 83, 2006.

[39] M. Gillick, I. Robertson, and J. Joshi, "Miniaturised Two-Stage Balanced Ku-band CPW MMIC Amplifier Using Impedance Transforming Couplers," *IEEE Electronic Letters*, vol. 29, no. 8, pp. 670 – 671, 1993.

[40] G. Gonzalez, *Microwave Transitor Amplifiers: Analysis and Design*. Prentice-Hall, Inc., 1997.

[41] A. Grebennikov, *RF and Microwave Power Amplifier Design*. New York: McGraw Hill, 2005.

[42] B. Green, K. Chu, E. Chumbes, J. Smart, J. Shealy, and L. Eastman, "The Effect of Surface Passivation on the Microwave Characteristics of Undoped AlGaN/GaN HEMTs," *IEEE Electron Device Letters*, vol. 21, no. 6, pp. 268 – 271, June 2000.

[43] B. Green, V. Tilak, V. Kaper, J. Smart, J. Shealy, and L. Eastman, "Microwave Power Limits of AlGaN/GaN HEMTs under Pulsed-Bias Conditions," *IEEE Transactions on Microwave Theory and Techniques*, vol. 51, no. 2, pp. 618 – 623, 2003.

[44] K. Gupta, R. Garg, and I. Bahl, *Microstrip Lines and Slotlines*. Boston, London: Artech House, 1979.

[45] R. Hoffmann, *Integrierte Mikrowellenschaltungen: Elektrische Grundlagen, Dimensionierung, technische Ausführung, Technologien*. Berlin, Heidelberg, New York: Springer Verlag, 1983.

[46] W. Holpp and C. Worning, "New Electronically Scanned Array Radars for Airborne Applications," in *Asia-Pacific Microwave Conference*, June 2007, pp. 1 – 5.

[47] T. Huet, J. Gruenenpuett, Z. Ouarch, D. Bouw, V. Serru, M. Camiade, C. Chang, and P. Chaumas, "A 8W High Efficiency X-band Power pHEMT Amplifier," in *Proceedings of the $3^{rd}$ European Microwave Integrated Circuits Conference, EuMIC 2008*, Amsterdam, Netherlands, Oct. 2008, pp. 289 – 292.

[48] B. Jacobs, "Towards Integrated AlGaN/GaN Based X-Band High-Power Amplifiers," Ph.D. dissertation, Technical University Eindhoven, Netherlands, July 2004.

[49] K.-I. Jeon, Y.-S. Kwon, and S.-C. Hong, "Input Harmonics Control Using non-linear Capacitor in GaAs FET Power Amplifier," in *IEEE MTT-S International Microwave Symposium Digest*, vol. 2, 1997, pp. 817 – 820.

[50] H. Klockenhoff, R. Behtash, J. Wurfl, W. Heinrich, and G. Trankle, "A Compact 16 Watt X-Band GaN-MMIC Power Amplifier," in *IEEE MTT-S Proceedings of the International Microwave Symposium Digest*, May 2006, pp. 1846 – 1849.

[51] J. Komiak, W. Kong, and K. Nichols, "High Efficiency Wideband 6 to 18 GHz PHEMT Power Amplifier MMIC," in *IEEE MTT-S International Microwave Symposium Digest*, June 2002, pp. 905 – 907.

[52] B. Kopp and D. Heston, "High-Efficiency 5-Watt Power Amplifier with Harmonic Tuning," in *IEEE MTT-S International Microwave Symposium Digest*, vol. 2, 1988, pp. 839 – 842.

[53] W. Kopp and S. Pritchett, "High Efficiency Power Amplification for Microwave and Millimeter Frequencies," in *IEEE MTT-S International Microwave Symposium Digest*, vol. 3, 1989, pp. 857 – 858.

[54] ——, "Harmonic Tuning of Power FETs at X-band," in *IEEE MTT-S International Microwave Symposium Digest*, vol. 3, 1990, pp. 989 – 992.

[55] J. Kühn, M. Musser, F. van Raay, R. Kiefer, M. Seelmann-Eggebert, M. Mikulla, R. Quay, T. Rödle, and O. Ambacher, "Design and realization of GaN RF-devices and circuits from 1 to 30 GHz," *International Journal of Microwave and Wireless Technologies*, vol. 2, no. 1, pp. 115 – 120, 2010.

[56] J. Kühn, F. van Raay, R. Quay, R. Kiefer, , M. S.-E. W. Bronner, M. Schlechtweg, M. Mikulla, O. Ambacher, and M. Thumm, "Balanced Microstrip AlGaN/GaN HEMT Power Amplifier MMIC for X-Band Applications," in *Proceedings of the* $3^{rd}$ *European Microwave Integrated Circuits Conference, EuMIC 2008*, Amsterdam, Netherlands, Oct. 2008, pp. 95 – 98.

[57] J. Kühn, F. van Raay, R. Quay, R. Kiefer, D. Peschel, M. Mikulla, M. Seelmann-Eggebert, W. Bronner, M. Schlechtweg, O. Ambacher, and M. Thumm, "Design of X-Band GaN MMICs Using Field Plates," in *Proceedings of the* $4^{th}$ *European Microwave Integrated Circuits Conference, EuMIC 2009*, Rome, Italy, Sept./Oct. 2009, pp. 33 – 36.

[58] J. Kühn, F. van Raay, R. Quay, R. Kiefer, M. Seelmann-Eggebert, T. Maier, R. Stibal, M. Mikulla, W. Bronner, M. Schlechtweg, O. Ambacher, and M. Thumm, "Design of Highly-Efficient GaN X-Band-Power-Amplifier MMICs," in *IEEE MTT-S Proceedings*

*of the International Microwave Symposium Digest*, Boston, MA, USA, June 2009, pp. 661 – 664.

[59] J. Kühn, F. van Raay, R. Quay, R. Kiefer, M. Seelmann-Eggebert, M. Mikulla, W. Bronner, M. Schlechtweg, O. Ambacher, and M. Thumm, "Device and Design Optimization for AlGaN/GaN X-Band-Power-Amplifiers with High Efficiency," *Journal of Infrared Millimeter and Terahertz Waves*, vol. 31, no. 3, pp. 367 – 379, 2010.

[60] J. Kühn, P. Waltereit, F. van Raay, R. Aidam, R. Quay, O. Ambacher, and M. Thumm, "Harmonic Termination of AlGaN/GaN/(Al)GaN-Single- and Double-Heterojunction HEMTs," in *German Microwave Conference, 2010*, Berlin, Germany, Mar. 2010, pp. 122 – 125.

[61] J. Lange, "Interdigitated Stripline Quadrature Hybrid," *IEEE Transactions on Microwave Theory and Techniques*, pp. 1150 – 1151, Dec. 1969.

[62] H. Le, Y. Shih, V. Hwang, T. Chi, K. Kasel, and D. Wang, "An X-band High-Efficiency MMIC Power Amplifier with 20-dB Return Losses," *IEEE Journal of Solid-State Circuits*, vol. 26, no. 10, Oct. 1991.

[63] Y.-S. Lin, Y.-W. Lain, and S. Hsu, "AlGaN/GaN HEMTs With Low Leakage Current and High On/Off Current Ratio," *IEEE Electron Device Letters*, vol. 31, no. 2, pp. 102 – 104, 2010.

[64] M. Ludwig, C. Buck, C. Mangenot, and M. Suess, "Impact of New Technologies on Future Space-borne Radar Design," in *IEEE International Geoscience and Remote Sensing Symposium*, vol. 3, July 2003, pp. 2137 – 2139.

[65] M. Maeda, H. Masato, H. Takehara, M. Nakamura, S. Morimoto, H. Fujimoto, Y. Ota, and O. Ishikawa, "Source Second-Harmonic Control for High Efficiency Power Amplifiers," *IEEE Transactions on Microwave Theory and Techniques*, vol. 43, no. 12/2, pp. 2952 – 2957, 1995.

[66] G. L. Matthaei, L. Young, and E. M. T. Jones, *Microwave Filters, Impedance-Matching Networks and coupling Structures*. New York, Toronto, London: McGraw-Hill Book Company, 1964.

[67] P. McGovern, J. Benedikt, P. Tasker, J. Powell, K. Hilton, J. Glasper, R. Balmer, T. Martin, and M. Uren, "Analysis of DC-RF Dispersion in AlGaN/GaN HFETs Using Pulsed I-V and Time-Domain Waveform Measurements," in *IEEE MTT-S International Microwave Symposium Digest*, 2005, pp. 503 – 506.

[68] D. Meharry, R. Lender, K. Chu, L. Gunter, and K. Beech, "Multi-Watt Wideband MMICs in GaN and GaAs," in *IEEE MTT-S International Microwave Symposium Digest*, June 2007, pp. 631 – 634.

[69] H.-J. Michel, *Zweitor-Analyse mit Leistungswellen.* Stuttgart, Germany: B. G. Teubner, 1981.

[70] U. Mishra, L. Shen, T. Kazior, and Y. Wu, "GaN-Based RF Power Devices and Amplifiers," *IEEE Proceedings*, vol. 96, no. 2, pp. 287 – 305, Feb. 2008.

[71] J. Moon, D. Wong, M. Antcliffe, P. Hashimoto, M. Hu, P. Willadsen, M. Micovic, H. Moyer, A. Kurdoghlian, P. MacDonald, M. Wetzel, and R. Bowen, "High PAE 1 mm AlGaN/GaN HEMTs for 20 W and 43% PAE X-band MMIC Amplifiers," in *IEDM Technical Digest*, Dec. 2006, pp. 1 – 2.

[72] M. Ohtomo, "Stability Analysis and Numerical Simulation of Multidevice Amplifiers," *IEEE Transactions on Microwave Theory and Techniques*, vol. 41, no. 6/7, pp. 983 – 991, June/July 1993.

[73] ——, "Proviso on the Unconditional Stability Criteria for Linear Twoport," *IEEE Transactions on Microwave Theory and Techniques*, vol. 43, no. 5, pp. 1197 – 1200, May 1995.

[74] V. Palankovski, S. Vitanov, and R. Quay, "Field-Plate Optimization of AlGaN/GaN HEMTs," in *IEEE Compound Semiconductor Integrated Circuit Symposium*, 2006, pp. 107 – 110.

[75] J. W. Palmour, S. T. Sheppard, R. P. Smith, S. T. Allen, W. L. Pribble, T. J. Smith, Z. Ring, J. J. Sumakaris, A. W. Saxler, and J. W. Milligan, "Wide Bandgap Semiconductor Devices and MMICs for RF Power Applications," in *IEDM Technical Digest*, Dec. 2001, pp. 385 – 388.

[76] A. Pantellini, M. Peroni, A. Nanni, A. Cetronio, A. Bettidi, and E. Giovine, "Gate Technology and Substrate Property Influence on GaN HEMT Switch Device Performance," in *Proceedings of the* $4^{th}$ *European Microwave Integrated Circuits Conference, EuMIC 2009*, Rome, Italy, Sept./Oct. 2009, pp. 140 – 143.

[77] Y. Pei, R. Chu, L. Shen, N. Fichtenbaum, Z. Chen, D. Brown, S. Keller, S. Denbaars, and U. Mishra, "Effect of Al Composition and Gate Recess on Power Performance of AlGaN/GaN High-Electron Mobility Transistors," *IEEE Electron Device Letters*, vol. 29, no. 4, pp. 300 – 302, 2008.

[78] R. Pengelly, *Microwave Field-Effect Transistors Theory, Design and Applications (3rd ed.)*. Atlanta: Noble Publishing, 1994.

[79] S. Piotrowicz, E. Morvan, R. Aubry, S. Bansropun, T. Bouvet, E. Chartier, T. Dean, O. Drisse, C. Dua, D. Floriot, M. di-Forte Poisson, Y. Gourdel, A. Hydesa, J. Jacquet, O. Jardel, D. Lancereau, J. M. Lean, G. Lecoustre, A. Martin, Z. Ouarch, T. Reveyrand, M. Richard, N. Sarazin, D. Thenot, and S. Delage, "State of the Art 58 W, 38 % PAE X-Band AlGaN/GaN HEMTs Microstrip MMIC Amplifiers," in *IEEE Compound Semiconductor Integrated Circuit Symposium Digest*, 2008, pp. 1 – 4.

[80] S. Piotrowicz, Z. Ouarch, E. Chartier, R. Aubry, G.Callet, D. Floriot, J.C.Jacquet, O. Jardel, E. Morvan, T. Reveyrand, N. Sarazin, and S. Delage, "43W, 52% PAE X-Band AlGaN/GaN HEMTs MMIC Amplifiers," in *IEEE MTT-S Proceedings of the International Microwave Symposium Digest*, Piscataway, NJ, USA, May 2010, pp. 505 – 508.

[81] A. Platzker, A. Palevsky, S. Nash, W. Struble, and Y. Tajima, "Characterization of GaAs Devices by a Versatile Pulsed I-V Measurement System," in *Microwave Symposium Digest, 1990., IEEE MTT-S International*, vol. 3, 1990, pp. 1137 – 1140.

[82] N. Pothecary, *Feedforward Linear Power Amplifiers*. Boston, London: Artech House, 1999.

[83] R. Quay, "Analysis and Simulation of High Electron Mobility Transistors," Ph.D. dissertation, Technische Universität Wien, July 2001.

[84] ——, *Gallium Nitride Electronics.* Berlin: Springer Verlag, 2008.

[85] R. Quay and M. Mikulla, "Gallium Nitride RF-Devices: An Overview on the Development Activities in Europe," in *IEEE MTT-S Proceedings of the International Microwave Symposium Digest*, Piscataway, NJ, USA, May 2010, pp. 1234 – 1237.

[86] R. Quay, M. Mikulla, P. Waltereit, F. van Raay, M. Dammann, J. Kühn, O. Ambacher, and P. Schuh, "Gallium Nitride MMICs for Future Reconnaissance and Imaging Applications," in *4th Security Research Conference 2009: Fraunhofer Symposium Future Security*, Karlsruhe, Germany, Sept. 2009.

[87] R. Quay, F. van Raay, J. Kühn, R. Kiefer, P. Waltereit, M. Musser, S. Maroldt, M. Dammann, M. Mikulla, and O. Ambacher, "L-to-Ka-Band AlGaN/GaN HEMT Power Amplifiers With High Efficiency," in *Semiconductor Conference Dresden, SCD 2009*, Dresden, Germany, Apr. 2009.

[88] R. Quay, F. van Raay, J. Kühn, R. Kiefer, P. Waltereit, M. Zorcic, M. Musser, W. Bronner, M. Dammann, M. Seelmann-Eggebert, M. Schlechtweg, M. Mikulla, O. Ambacher, J. Thorpe, K. Riepe, F. van Rijs, M. Saad, L. Harm, and T. Rödle, "Efficient AlGaN/GaN HEMT Power Amplifiers," in *Proceedings of the $3^{rd}$ European Microwave Integrated Circuits Conference, EuMIC 2008*, Amsterdam, Netherlands, Oct. 2008, pp. 87 – 90.

[89] F. Raab, "Maximum Efficiency and Output of Class-F Power Amplifiers," *IEEE Transactions on Microwave Theory and Techniques*, vol. 49, no. 6, pp. 1162 – 1166, 2001.

[90] F. Raab, P. Asbeck, S. Cripps, P. Kenington, Z. Popovic, N. Pothecary, J. Sevic, and N. Sokal, "Power Amplifiers and Transmitters for RF and Microwave," *IEEE Transactions on Microwave Theory and Techniques*, vol. 50, no. 3, pp. 814 – 826, Mar. 2002.

[91] J. Reed and G. Wheeler, "A Method of Analysis of Symmetrical Four-Port Networks," *IRE Transactions on Microwave Theory and Techniques*, vol. 4, no. 4, pp. 246 – 252, 1956.

[92] C. Roff, A. Sheikh, J. Benedikt, P. Tasker, K. Hilton, J. Maclean, D. Hayes, M. Uren, and T. Martin, "Optimising AIGaN/GaN

HFET Designs for High Efficiency," in *Proceedings of the* $2^{nd}$ *European Microwave Integrated Circuits Conference, EuMIC 2007*, Munich, Germany, Oct. 2007, pp. 165 – 168.

[93] J. Rollett, "Stability and power-gain invariants of linear two-ports," *IRE Transactions on Circuit Theory*, vol. CT-9, pp. 29 – 32, Mar. 1962.

[94] M. Rosker, "The Present State of the Art of Wide-Bandgap Semiconductors and Their Future," in *IEEE Radio Frequency Integrated Circuits Symposium*, June 2007, pp. 159 – 162.

[95] ——, "The DARPA Wide Band Gap Semiconductors for RF Applications (WBGS-RF) Program: Phase II Results," in *CS ManTech*, 2009, p. 1.2.

[96] K. Russell, "Microwave Power Combining Techniques," *IEEE Transactions on Microwave Theory and Techniques*, vol. 27, no. 5, pp. 472 – 478, May 1979.

[97] W. Saito, Y. Kakiuchi, T. Nitta, Y. Saito, T. Noda, H. Fujimoto, A. Yoshioka, T. Ohno, and M. Yamaguchi, "Field-Plate Structure Dependence of Current Collapse Phenomena in High-Voltage GaN-HEMTs," *IEEE Electron Device Letters*, vol. 31, no. 7, pp. 659 – 661, July 2010.

[98] W. Saito, M. Kuraguchi, Y. Takada, K. Tsuda, I. Omura, and T. Ogura, "Influence of Surface Defect Charge at AlGaN-GaN-HEMT upon Schottky Gate Leakage Current and Breakdown Voltage," *IEEE Transactions on Electron Devices*, vol. 52, no. 2, pp. 159 – 164, 2005.

[99] F. Scappaviva, R. Cignani, C. Florian, G. Vannini, F. Filicori, and M. Feudale, "10 Watt High Efficiency GaAs MMIC Power Amplifier for Space Applications," in *Proceedings of the 3rd European Microwave Integrated Circuits Conference, EuMIC 2008*, Amsterdam, Netherlands, Oct. 2008, pp. 562 – 565.

[100] P. Schuh, R. Leberer, H. Sledzik, M. Oppermann, B. Adelseck, H. Brugger, R. Behtash, H. Leier, R. Quay, and R. Kiefer, "20W GaN HPAs for Next Generation X-Band T/R-Modules," in *IEEE MTT-S International Microwave Symposium Digest*, June 2006, pp. 726 – 729.

[101] P. Schuh, H. Sledzik, R. Reber, A. Fleckenstein, R. Leberer, M. Oppermann, R. Quay, F. van Raay, M. Seelmann-Eggebert, R. Kiefer, and M. Mikulla, "GaN MMIC based T/R-Module Front-End for X-Band Applications," in *Proceedings of the 3rd European Microwave Integrated Circuits Conference, EuMIC 2008*, Amsterdam, Netherlands, Oct. 2008, pp. 274 – 277.

[102] ——, "X-Band T/R-module front-end based on GaN MMICs," *International Journal of Microwave and Wireless Technologies*, vol. 1, no. 4, pp. 387 – 394, 2009.

[103] O. Schuler, H. Fourre, R. Fauquembergue, and A. Cappy, "Influence of Parasitic Capacitances on the Performance of Passivated InAlAs/InGaAs HEMTs in the Millimeter Wave Range," in *IPRM International Conference on Indium Phosphide and Related Materials*, 1996, pp. 646 – 649.

[104] F. Schwierz and O. Ambacher, "Recent Advances in GaN HEMT Development," in *IEEE International Symposium on Electron Devices for Microwave and Optoelectronic Applications*, Nov. 2003, pp. 204 – 209.

[105] F. Sechi and M. Bujatti, *Solid-State Microwave High-Power Amplifiers*. Boston, London: Artech House, 2009.

[106] M. Seelmann-Eggebert, T. Merkle, F. van Raay, R. Quay, and M. Schlechtweg, "A Systematic State-Space Approach to Large-Signal Transistor Modeling," *IEEE Transactions on Microwave Theory and Techniques*, vol. 55, no. 2, pp. 195 – 205, Feb. 2007.

[107] S. Seo, D. Pavlidis, and J. Moon, "Wideband balanced AlGaN/GaN HEMT MMIC low noise amplifier," *IEEE Electronic Letters*, vol. 41, no. 16, pp. 37 – 38, Aug. 2005.

[108] L. Shen, T. Palacios, C. Poblenz, A. Corrion, A. Chakraborty, N. Fichtenbaum, S. Keller, S. Denbaars, J. Speck, and U. Mishra, "Unpassivated High Power Deeply Recessed GaN HEMTs with Fluorine-Plasma Surface Treatment," *IEEE Electron Device Letters*, vol. 27, no. 4, pp. 214 – 216, Apr. 2006.

[109] J. Shumaker, M. Ohoka, and N. Ui, "Design of Power Amplifiers Using High Breakdown GaN HEMT Devices," in *IEEE International Conference on Microwaves, Communications, Antennas and Electronic Systems*, 2008, pp. 1 – 5.

[110] R. Singh and C. Snowden, "Small-Signal Characterization of Microwave and Millimeter-Wave HEMT's based on a Physical Model," *IEEE Transactions on Microwave Theory and Techniques*, vol. 44, no. 1, pp. 114 – 121, jan 1996.

[111] H. Sledzik and P. Schuh, "Personal Message," 2010, EADS, Ulm.

[112] M. Spirito, L. de Vreede, L. Nanver, J. Mueller, and J. Burghartz, "Low-loss Passives for 2nd-Harmonic Termination Control in Power Amplifiers for Mobile Applications," in *2003 Topical Meeting on Silicon Monolithic Integrated Circuits in RF Systems*, 2003, pp. 49 – 52.

[113] M. Spirito, L. de Vreede, L. Nanver, S. Weber, and J. Burghartz, "Power Amplifier PAE and Ruggedness Optimization by Second Harmonic Control," in *Proceedings of the 2002 Bipolar/ BiCMOS Circuits and Technology Meeting*, 2002, pp. 173 – 176.

[114] C. Tang, M. Chen, and C. Tsai, "Miniaturization of Microstrip Branch-Line Coupler With Dual Transmission Lines," *IEEE Microwave and Wireless Components Letters*, vol. 18, no. 3, pp. 185 – 187, 2008.

[115] C. Tangsheng, Z. Bin, R. Chunjiang, J. Gang, Z. Weibin, C. Chen, S. Kai, and Y. Naibin, "X-band microstrip algan/gan hemt power mmics," in *IEEE Proceedings of Compound Semiconductor Integrated Circuits Symposium, CSICS 2008*, Monterey, Ca, USA, Oct. 2008, pp. 1 – 4.

[116] D. Teeter, A. Platzker, and R. Bourque, "A Compact Network for Eliminating Parametric Oscillations in High Power MMIC Amplifiers," in *IEEE MTT-S International Microwave Symposium Digest*, vol. 3, 1999, pp. 967 – 970.

[117] M. Thumm, *Hoch- und Höchstfrequenz-Halbleiterschaltungen (in German)*. Skriptum zur Vorlesung: Universität Karlsruhe (TH), Germany, 2002/2003.

[118] R. Trew, G. Bilbro, W. Kuang, Y. Liu, and H. Yin, "Microwave AlGaN/GaN HFETs," *IEEE Electron Device Letters*, vol. 6, no. 1, pp. 56 – 66, Mar. 2005.

[119] H. van Bezouwen, H. Feldle, and W. Holpp, "Status and Trends in AESA-based Radar," in *IEEE MTT-S Proceedings of the International Microwave Symposium Digest*, Piscataway, NJ, USA, May 2010, pp. 526 – 529.

[120] F. van den Bogaart and T. de Hek, "First-Pass Design Strategy for High-Power Amplifiers at X-band," in *IEE Tutorial Colloquium on Design of RFIC's and MMIC's*, vol. 391, 1997, pp. 8/1 – 8/6.

[121] G. van der Bent, A. de Hek, A. Bessemoulin, and F. van Vliet, "Low-Cost High-Efficient 10-Watt X-band High-Power Amplifier," in *IEEE International Conference on Microwaves, Communications, Antennas and Electronic Systems*, Nov. 2009, pp. 1 – 6.

[122] F. van Raay, "Personal Message," 2010, Fraunhofer IAF, Freiburg.

[123] F. van Raay, R. Quay, R. Kiefer, F. Benkhelifa, B. Raynor, W. Pletschen, M. Kuri, H. Massler, S. Müller, M. Dammann, M. Mikulla, M. Schlechtweg, and G. Weimann, "A Coplanar X-band AlGaN/GaN Power Amplifier MMIC on s.i. SiC Substrate," *IEEE Microwave and Wireless Components Letters*, vol. 15, no. 7, pp. 460 – 462, 2005.

[124] F. van Raay, R. Quay, R. Kiefer, W. Bronner, M. Seelmann-Eggebert, M. Mikulla, M. Schlechtweg, and G. Weimann, "X-Band High-Power Microstrip AlGaN/ GaN HEMT Amplifier MMICs," in *IEEE MTT-S International Microwave Symposium Digest*, 2006, pp. 1368 – 1371.

[125] F. van Raay, R. Quay, R. Kiefer, W. Fehrenbach, W. Bronner, M. Kuri, F. Benkhelifa, H. Massler, S. Müller, M. Mikulla, M. Schlechtweg, and G. Weimann, "A Microstrip X-Band AlGaN/GaN Power Amplifier MMIC on s.i. SiC Substrate," in *European Gallium Arsenide and Other Semiconductor Application Symposium*, Paris, France, Oct. 2005, pp. 233 – 236.

[126] F. van Raay, R. Quay, R. Kiefer, H. Walcher, O. Kappeler, M. Seelmann-Eggebert, S. Müller, M. Schlechtweg, and G. Weimann, "High Power/High Bandwidth GaN MMICs and Hybrid Amplifiers: Design and Characterization," in *European*

*Gallium Arsenide and Other Semiconductor Application Symposium*, Paris, France, Oct. 2005, pp. 373 – 376.

[127] G. D. Vendelin, A. M. Pavio, and U. L. Rohde, *Microwave Circuit Design Using Linear and Nonlinear Techniques*. John Wiley & Sons, Inc., 1990.

[128] G. Verzellesi, A. Mazzanti, C. Canali, G. M. A. Chini, and E. Zanoni, "Study on the Origin of Dc-to-RF Dispersion Effects in GaAs- and GaN-based Heterostructure FETs," in *IEEE Proceedings GaAs Reliability Workshop*, 2003, pp. 155 – 156.

[129] R. Vetury, N. Zhang, S. Keller, and U. Mishra, "The Impact of Surface States on the DC and RF Characteristics of AlGaN/GaN HFETs," *IEEE Transactions on Electron Devices*, vol. 48, no. 3, pp. 560 – 566, 2001.

[130] J. Walker, *High-Power GaAs FET Amplifiers*. Boston, London: Artech House, 1993.

[131] P. Waltereit, W. Bronner, R. Kiefer, R. Quay, J. Kühn, F. van Raay, M. Dammann, S. Müller, M. Mikulla, and O. Ambacher, "High Efficiency and Low Leakage AlGaN/GaN HEMTs for a Reproducible and Reliable X-band MMIC Technology," in *CS Mantech 2010*, Portland, Oregon, USA, May 2010.

[132] P. Waltereit, W. Bronner, R. Quay, M. Dammann, R. Kiefer, S. Müller, M. Musser, J. Kühn, F. van Raay, M. Seelmann-Eggebert, M. Mikulla, O. Ambacher, F. van Rijs, T. Rödle, and K. Riepe, "GaN HEMT and MMIC Development at Fraunhofer IAF: Performance and Reliability," *Physica status solidi (a)*, vol. 206, no. 6, pp. 1215 – 1220, Mar. 2009.

[133] P. Waltereit, W. Bronner, R. Quay, M. Dammann, R. Kiefer, W. Pletschen, S. Müller, R. Aidam, H. Menner, L. Kirste, K. Köhler, M. Mikulla, and O. Ambacher, "AlGaN/GaN Epitaxy and Technology," *International Journal of Microwave and Wireless Technologie 2*, vol. 1, pp. 3 – 11, 2010.

[134] P. Waltereit, S. Müller, K. Bellmann, C. Buchheim, R. Goldhahn, K. Köhler, L. Kirste, M. Baeumler, M. Dammann, W. Bronner, R. Quay, and O. Ambacher, "Impact of GaN cap Thickness on optical, Electrical, and Device Properties in AlGaN/GaN

High Electron Mobility Transistor Structures," *Journal of Applied Physics*, vol. 106, no. 23, pp. 023535 – 1 – 7, 2009.

[135] S. Watanabe, S. Takatuka, K. Takagi, H. Kuroda, and Y. Oda, "Simulation and Experimental Results of Source Harmonic Tuning on Linearity of Power GaAs FET under Class AB Operation," in *IEEE MTT-S International Microwave Symposium Digest*, vol. 3, 1996, pp. 1771 – 1774.

[136] R. Waterman, W. Fabian, R. Pucel, Y. Tajima, and J. Vorhaus, "GaAs Monolithic Lange and Wilkinson Couplers," *IEEE Transactions on Electron Devices*, vol. 28, no. 2, Feb. 1981.

[137] P. White, "Effect of Input Harmonic Terminations on High Efficiency Class-B and Class-F Operation of PHEMT Devices," in *IEEE MTT-S International Microwave Symposium Digest*, vol. 3, 1998, pp. 1611 – 1614.

[138] P. White and T. O'Leary, "A 50% Efficiency 8 W C-band PHEMT Power MMIC Amplifier," in *IEEE Gallium Arsenide Integrated Circuit (GaAs IC) Symposium*, 1995, pp. 277 – 280.

[139] Y. Wu, M. Moore, A. Saxler, T. Wisleder, and P. Parikh, "40-W/mm Double Field-plated GaN HEMTs," in *64th Device Research Conference*, 2006, pp. 151 – 152.

[140] Y. Wu, A. Saxler, M. Moore, R. Smith, S. Sheppard, P. Chavarkar, T. Wisleder, U. Mishra, and P. Parikh, "30-W/mm GaN HEMTs by Field Plate Optimization," *IEEE Electron Device Letters*, vol. 25, no. 3, pp. 117 – 119, Mar. 2004.

[141] Y. Wu, A. Saxler, M. Moore, T. Wisleder, U. Mishra, and P. Parikh, "Field-plated GaN HEMTs and Amplifiers," in *IEEE Compound Semiconductor Integrated Circuit Symposium*, 2005, pp. 170 – 172.

[142] H. Xu, C. Sanabria, A. Chini, W. Yun, S. Heikman, S. Keller, U. Mishra, and R. York, "A new Field-Plated GaN HEMT Structure with Improved Power and Noise Performance," in *IEEE Lester Eastman Conference on High Performance Devices*, 2004, pp. 186 – 191.

[143] A. Zhang, L. Rowland, E. Kaminsky, J. Kretchmer, R. Beaupre, J. Garrett, and J. Tucker, "Microwave Power SiC MESFETs and GaN HEMTs," in *IEEE Lester Eastman Conference on High Performance Devices*, 2002, pp. 181 – 185.

[144] J. Zolper, "Wide Bandgap Semiconductor Microwave Technologies: From Pro-mise to Practice," in *IEDM International Electron Devices Meeting*, 1999, pp. 389 – 392.

# List of Personal Publications


[P-1] P. Waltereit, W. Bronner, R. Kiefer, R. Quay, **J. Kühn**, F. van Raay, M. Dammann, S. Müller, M. Mikulla, and O. Ambacher, "High Efficiency and Low Leakage AlGaN/GaN HEMTs for a Reproducible and Reliable X-band MMIC Technology," in *CS Mantech 2010*, Portland, Oregon, USA, May 2010.

[P-2] **J. Kühn**, M. Musser, F. van Raay, R. Kiefer, M. Seelmann-Eggebert, M. Mikulla, R. Quay, T. Rödle, and O. Ambacher, "Design and realization of GaN RF-devices and circuits from 1 to 30 GHz," *Cambridge Journals Online*, vol. 2, no. 1, pp. 115 – 120, 2010.

[P-3] **J. Kühn**, F. van Raay, R. Quay, R. Kiefer, M. Mikulla, M. Seelmann-Eggebert, W. Bronner, M. Schlechtweg, O. Ambacher, and M. Thumm, "Device and Design Optimization for AlGaN/GaN X-Band-Power-Amplifiers with High Efficiency," *Journal of Infrared Millimeter and Terahertz Waves*, vol. 31, no. 3, pp. 367 – 379, 2010.

[P-4] **J. Kühn**, P. Waltereit, F. van Raay, R. Aidam, R. Quay, O. Ambacher, and M. Thumm, "Harmonic Termination of AlGaN/GaN/(Al)GaN-Single- and Double-Heterojunction HEMTs," in *German Microwave Conference 2010*, Berlin, Germany, Mar. 2010, pp. 122 – 125.

[P-5] P. Waltereit, W. Bronner, R. Quay, M. Dammann, R. Kiefer, S. Müller, M. Musser, **J. Kühn**, F. van Raay, M. Seelmann-Eggebert, M. Mikulla, O. Ambacher, F. van Rijs, T. Rödle, and K. Riepe, "GaN HEMT and MMIC Development at Fraunhofer IAF: Performance and Reliability," *Physica status solidi (a)*, vol. 206, no. 6, pp. 1215 – 1220, Mar. 2009.

[P-6] R. Quay, M. Mikulla, P. Waltereit, F. van Raay, M. Dammann, **J. Kühn**, O. Ambacher, and P. Schuh, "Gallium Nitride MMICs for Future Reconnaissance and Imaging Applications," in *4th Security Research Conference 2009: Fraunhofer Symposium Future Security*, Karlsruhe, Germany, Sept. 2009.

[P-7] R. Quay, F. van Raay, **J. Kühn**, R. Kiefer, P. Waltereit, M. Musser, S. Maroldt, M. Dammann, M. Mikulla, and O. Ambacher, "L-to-Ka-Band AlGaN/GaN HEMT Power Amplifiers With High Efficiency," in *Semiconductor Conference Dresden, SCD 2009*, Dresden, Germany, Apr. 2009.

[P-8] **J. Kühn**, F. van Raay, R. Quay, R. Kiefer, D. Peschel, M. Mikulla, M. Seelmann-Eggebert, W. Bronner, M. Schlechtweg, O. Ambacher, and M. Thumm, "Design of X-Band GaN MMICs Using Field Plates," in *Proceedings of the* $4^{th}$ *European Microwave Integrated Circuits Conference, EuMIC*, Rome, Italy, Sept. 2009, pp. 33 – 36.

[P-9] **J. Kühn**, F. van Raay, R. Quay, R. Kiefer, T. Maier, R. Stibal, M. Mikulla, M. Seelmann-Eggebert, W. Bronner, M. Schlechtweg, O. Ambacher, and M. Thumm, "Design of Highly-Efficient GaN X-Band-Power-Amplifier MMICs,"in *IEEE MTT-S Proceedings of the International Microwave Symposium Digest*, Boston, MA, USA, June 2009, pp. 661 – 664.

[P-10] R. Quay, F. van Raay, **J. Kühn**, R. Kiefer, P. Waltereit, M. Zorcic, M. Musser, W. Bronner, M. Dammann, M. Seelmann-Eggebert, M. Schlechtweg, M. Mikulla, O. Ambacher, J. Thorpe, K. Riepe, F. van Rijs, M. Saad, L. Harm, and T. Rödle, "Efficient AlGaN/GaN HEMT Power Amplifiers," in *Proceedings of the* $3^{rd}$ *European Microwave Integrated Circuits Conference, EuMIC 2008*, Amsterdam, Netherlands, Oct. 2008, pp. 87 – 90.

[P-11] **J. Kühn**, F. van Raay, R. Quay, R. Kiefer, W. Bronner, M. Seelmann-Eggebert, M. Schlechtweg, M. Mikulla, O. Ambacher, and M. Thumm, "Balanced Microstrip AlGaN/GaN HEMT Pow-er Amplifier MMIC for X-Band Applications," in *Proceedings of the* $3^{rd}$ *European Microwave Integrated Circuits Conference, EuMIC 2008*, Amsterdam, Netherlands, Oct. 2008, pp. 95 – 98.

# List of Symbols

| | |
|---|---|
| $\alpha$ | Conduction angle |
| $\Gamma_{\mathrm{ii}}$ | Reflection coefficient |
| $\Gamma_{\mathrm{in}}$ | Input impedance of HEMT device |
| $\Gamma_{\mathrm{L}}$ | Load (output) reflection coefficient |
| $\Gamma_{\mathrm{L,\,opt}}$ | Optimum load impedance |
| $\Gamma_{\mathrm{L,\,Pout}}$ | Optimum load impedance for maximum $P_{\mathrm{out}}$ |
| $\Gamma_{\mathrm{L,\,PAE}}$ | Optimum load impedance for maximum PAE |
| $\Gamma_{\mathrm{S}}$ | Source (input) reflection coefficient |
| $\Gamma_{\mathrm{S,\,Pout}}$ | Optimum input impedance for maximum $P_{\mathrm{out}}$ |
| $\Gamma_{\mathrm{S,\,PAE}}$ | Optimum input impedance for maximum PAE |
| $\varepsilon_{\mathrm{r}}$ | Relative dielectric constant |
| $\varepsilon_{\mathrm{r,\,eff}}$ | Effective relative dielectric constant |
| $\eta_{\mathrm{D}}$ | Drain efficiency |
| $\Theta$ | Thermal conductivity |
| $\lambda$ | Wavelength |
| $\mu$ | Electron mobility |
| $\mu_{\mathrm{r}}$ | Relative magnetic constant |
| $\tau$ | Channel transit time |
| $v_{\mathrm{sat}}$ | Saturated electron velocity |
| $\omega$ | Angular frequency |
| $A_{\mathrm{s}}$ | Amplitude of incoming wave $\mathrm{a_i}$ at source |
| $\mathrm{a_i}$ | Incoming wave |
| $\mathrm{b_i}$ | outgoing wave |
| $\mathrm{BV_{GD}}$ | Breakdown voltage for the gate-drain diode |
| $\mathrm{BV_{GS}}$ | Breakdown voltage for the gate-source diode |
| $BW_{\mathrm{rel}}$ | Relative Bandwidth |
| $C_i$ | Capacitor |

| | |
|---|---|
| $c_0$ | Speed of light in vacuum |
| $C_{\mathrm{add}}(f_0)$ | Additional capacitance at $f_0$ |
| $C_{\mathrm{ds}}$ | Drain-source capacitance |
| $C_{\mathrm{eff,\,out}}$ | Effective output capacitance |
| $C_{\mathrm{eff,\,out\,HT}}$ | Effective output capacitance including $C_{\mathrm{add}}$ |
| $C_{\mathrm{gd}}$ | Gate-drain capacitance |
| $C_{\mathrm{gs}}$ | Gate-source capacitance |
| $C_{\mathrm{pd}}$ | Parasitic drain capacitance |
| $C_{\mathrm{pg}}$ | Parasitic gate capacitance |
| $C_{\mathrm{pgd}}$ | Parasitic gate-drain capacitance |
| $C_{\mathrm{P,\,add/DRV}}$ | Additional output power margin of the DRV-stage |
| $C_{\mathrm{P,\,add/PA}}$ | Additional output power margin of the PA-stage |
| $DE_{\mathrm{FET}}$ | Drain efficiency of a single FET cell |
| $\mathrm{E}_{\mathrm{br}}$ | Break-down field |
| $E_{\mathrm{g}}$ | Bandgap |
| $f$ | Frequency |
| $f_0$ | Fundamental/center frequency |
| $f_{\mathrm{c}}$ | Cut-off frequency |
| $f_{\mathrm{max}}$ | Maximum frequency of oscillation |
| $f_{\mathrm{T}}$ | Transit frequency |
| G | Gain |
| $G_{\mathrm{lin}}$ | Linear gain |
| $G_{\mathrm{P,\,DRV}}$ | Power gain of the DRV-stage |
| $G_{\mathrm{P,\,HPA}}$ | Power gain of the entire HPA |
| $G_{\mathrm{P,\,PA}}$ | Power gain of the PA-stage |
| $G_{\mathrm{T}}$ | Transducer gain |
| $g_{\mathrm{m}}$ | Intrinsic transconductance |
| $g_{\mathrm{m,\,peak}}$ | Maximum transconductance |
| $h$ | Substrate thickness |
| $I_{\mathrm{D}}$ | Drain current |
| $I_{\mathrm{DC}}$ | DC supply current |
| $I_{\mathrm{DS}}$ | Drain-source current |
| $I_{\mathrm{DS,\,int}}$ | Internal drain current |
| $I_{\mathrm{GS}}$ | Gate-source current |
| $I_{\mathrm{max}}$ | Maximum drain current |
| $I_{\mathrm{n}}$ | Drain current waveform of n-th harmonic |
| $I_{\mathrm{th}}$ | Threshold current |
| $k$ | Rollet's stability factor |
| $L_{comp}$ | Compensation inductance |

| | |
|---|---|
| $L_{comp,HT}$ | Compensation inductance depending on 2nd HT stub |
| $l_c$ | Mechanical length of coupler |
| $l_{FP}$ | Fieldplate length |
| $l_g$ | Gate length |
| $l_{gg}$ | Gate-to-Gate pitch |
| $L_d$ | Extrinsic gate inductance |
| $L_g$ | Extrinsic drain inductance |
| $L_i$ | Inductor |
| $L_{IMN}$ | Net insertion loss of the input matching network |
| $L_{ISMN}$ | Net insertion loss of the interstage matching network |
| $L_{net}$ | Net insertion loss |
| $L_{OMN}$ | Net insertion loss of the output matching Network |
| $L_s$ | Extrinsic source inductance |
| $N_{FET}$ | Number of transistors used in a transistor-stage |
| $N_{FET/DRV}$ | Number of transistors used in the DRV-stage |
| $N_{FET/PA}$ | Number of transistors used in the PA-stage |
| $N_{L,c}$ | Number of lines used in the coupler |
| PAE | Power-added efficiency |
| $PAE_{DRV}$ | Power-added efficiency of the DRV-stage |
| $PAE_{FET}$ | Power-added efficiency of a single FET cell |
| $PAE_{HPA}$ | Power-added efficiency of the High-Power Amplifier |
| $PAE_{PA}$ | Power-added efficiency of the PA-stage |
| $P_{DC}$ | DC supply power |
| $P_{DC,DRV}$ | DC supply power of the DRV-stage |
| $P_{DC,FET}$ | DC supply power of the FET |
| $P_{DC,PA}$ | DC supply power of the PA-stage |
| $P_{in}$ | Input power |
| $P_{in,DRV}$ | Input power of the DRV-stage |
| $P_{in,HPA}$ | Input power of the High-Power Amplifier |
| $P_{in,PA}$ | Input power of the PA-stage |
| $P_{out}$ | Output power |
| $P_{out,DRV}$ | RF Output power of the DRV-stage |
| $P_{out,HPA}$ | RF Output power of the High-Power Amplifier |
| $P_{out,PA}$ | RF Output power of the PA-stage |
| $P_{sat}$ | Saturated power |
| Q | Q-factor |
| $R_d$ | Extrinsic drain resistance |
| $R_{ds}$ | Drain-source resistance |
| $R_g$ | Extrinsic gate resistance |

| | |
|---|---|
| $R_{gd}$ | Gate-drain space charge region resistance |
| $R_{gs}$ | Gate-source resistance |
| $R_L$ | Load resistance |
| $R_{on}$ | On resistance |
| $R_s$ | Extrinxic source resistance |
| $r_s$ | Reflection coefficient at source |
| $r_{out}$ | Output impedance |
| $s_c$ | Gap between coupler lines |
| $S_{Coupler}$ | S-parameter matrix of a quadrature (90°) 3 dB coupler |
| $S_{ij}$ | Scattering Parameters (i,j = 1,2) |
| $S_{11}$ | Input scattering parameter |
| $S_{22}$ | Output scattering parameter |
| $S_{21}$ | Transmission scattering parameter |
| $S_{12}$ | Reverse transmission scattering parameter |
| $t$ | Metal thickness |
| $T_a$ | Ambient temperature |
| TGW | Total gate width |
| $TGW_{DRV}$ | Total gate width of the DRV-stage |
| $TGW_{HPA}$ | Total gate width of the High-Power Amplifier |
| $TGW_{PA}$ | Total gate width of the PA-stage |
| $V_{DC}$ | DC supply voltage |
| $V_{DS}$ | Drain-source voltage |
| $V_{DS,\,int}$ | Internal drain voltage |
| $V_{GS}$ | Gate-souce voltage |
| $V_k$ | Knee voltage |
| $V_{max}$ | Maximum Drain-source voltage |
| $V_{th}$ | Threshold voltage |
| $W_g$ | Gate width |
| $w_c$ | Single microstrip line width of coupler |
| $X_C$ | Reactance of $C_{eff,\,out}$ |
| $X_{C,\,add}(f_0)$ | Reactance of $C_{add}(f_0)$ |
| $Y_{ij}$ | Y-parameters |
| $Y_L$ | Load admittance corresponding to load reflection coefficient $\Gamma_L$ |
| $Z$ | Impedance |
| $Z_0$ | Reference impedance, usually 50 Ω |
| $Z_L$ | Load impedance |
| $Z_i$ | Image impedance |

# List of Abbreviations

| | |
|---|---|
| 2DEG | Two-Dimensional-Electron-Gas |
| ADS | Advanced Design System by Agilent Technologies |
| (Al)GaN | (Aluminum) Gallium Nitride |
| BW | Bandwidth |
| CAD | Computer Aided Design |
| CPW | Coplanar Waveguide |
| CW | Continuous Wave |
| DA | Driver amplifier |
| DRC | Design Rule Check |
| DC | Direct Current |
| DE | Drain Efficiency |
| DH-HEMT | Double-Heterojunction High Electron Mobility Transistor |
| DRV | Driver |
| DUT | Device under Test |
| FET | Field Effect Transistor |
| FIB | Focused Ion Beam |
| FMCW | Frequency Modulated Continuous Wave |
| FP | Field plate |
| GaAs | Gallium Arsenide |
| GaN | Gallium Nitride |
| GFP | Gate-connected field plate |
| GW | Gate-width |
| HEMT | High Electron Mobility Transistor |
| HF | High Frequency |
| HFSS | High Frequency Structural Simulator |
| HPA | High-Power Amplifier |

| | |
|---|---|
| HT | Harmonic Termination |
| IAF | Fraunhofer Institute for Applied Solid-State Physics |
| IC | Integrated Circuit |
| IMN | Input Matching Network |
| ISMN | Insterstage Matching Network |
| ISV | Individual Source Vias |
| I-V | Current Voltage |
| LNA | Low Noise Amplifier |
| LS | Large-Signal |
| LP | Load-Pull |
| MAG | Maximum Available Gain |
| MBE | Molecular Beam Epitaxy |
| MESFET | Metal Semiconductor Field Effect Transistor |
| MIC | Microwave Integrated Circuit |
| MIM | Metal-Insulator-Metal |
| MMIC | Monolithic Microwave Integrated Circuit |
| MN | Matching Network |
| MOCVD | Metal Organic Chemical Vapor Deposition |
| MSG | Maximum Stable Gain |
| MSL | Microstrip Line |
| NDPA | Non-Uniform Distributed Power Amplifier |
| NiCr | Nickel-Chromium |
| OMN | Output Matching Network |
| PA | Power Amplifier |
| pHEMT | Pseudomorphic High Electron Mobility Transistor |
| Radar | Radio Detection and Ranging |
| RF | Radio Frequency |
| SEM | Scanning Electron Microscope |
| SH | Shield |
| SH-HEMT | Single-Heterojunction High Electron Mobility Transistor |
| Si | Silicon |
| s.i. | Semi-insulating |
| SiC | Silicon Carbide |
| SiN | Silicon Nitride |
| S-parameter | Scattering Parameter |
| T/R-module | Transmit/receive-module |
| TGW | Total gate-width |
| VSWR | Voltage Standing Wave Ratio |

# Acknowledgments

This thesis is a result of my work as a scientist at the Fraunhofer Institute for Applied Solid-State Physics (IAF) in Freiburg, Germany.

First of all, I deeply acknowledge my advisor Prof. Dr. rer. nat. Manfred Thumm, Head of the Gyrotron Development and Microwave Technology Division, Institute for Technical Physics, at the Karlsruhe Institute of Technology (KIT), Germany. His encouragement, continuous support, and confidence in my capabilities enabled the successful realization of this thesis.

Moreover, I'm very grateful to Prof. Dr. rer. nat. Oliver Ambacher, Head of the Fraunhofer IAF, for giving me the opportunity to use the Institute's excellent facilities and environment without which this work could not have been realized as well as for taking the position of second reviewer.

I would like to thank Dr. Michael Schlechtweg, Head of the High Frequency Devices and Circuits Department at the Fraunhofer IAF, for his belief in my work and for his motivating and complaisant support.

My special thank is dedicated to Dr. Rüdiger Quay for his endless encouragement and for his extremely competent advice, constructive criticism, and support over the last few years. Moreover, I'm deeply grateful to Dr. Friedbert van Raay for putting tremendous amount of time and effort in mentoring my work by introducing me to the design of High-Power Amplifiers and finally for accurate proofreading. The various discussions with him helped me to solve many problems and gave me new ideas.

Furthermore, I like to thank Dr. Matthias Seelmann-Eggebert for providing very useful and accurate large-signal AlGaN/GaN HEMT models which are a profund and reliable base for the successful design of AlGaN/GaN HEMT High-Power Amplifiers. In addition, I thank Hermann Massler for the introduction to the world of RF measurement systems and the whole measurement group for their continuous and fast help with the measurements. Without the constant IAF AlGaN/GaN HEMT process development, this work would not have been possible. Therefore, I'm thankful to Dr. Michael Mikulla and Dr. Rudolf Kiefer representing the complete stuff members of the technology group.

I'm grateful to Brian Raynor for proofreading my work concerning the English language. For the excellent working atmosphere and the continuous encouragement during my time at the Fraunhofer IAF, I like to thank all my colleagues at IAF, especially my roommates Dr. Robert Makon and Ernst Weissbrodt. I also thank Hildegard Brucher for her everlasting good mood and endless friendliness.

In addition, I'm thankful to the IT group for the technical support, the simulation hardware and software environments both at the Fraunhofer IAF and also at home. The provided work station at home was very helpful, especially during the time of parental leave.

Last but not least, I would like to thank my family and parents. The love and support, especially concerning the care for our daughter Eva Marie, from all the Feiler, Ullrich, and Kühn families have carried me through my studies. I thank my husband Jochen for his patience and understanding. Due to his tireless efforts within all life situations he has kept me free for this doctoral thesis. His encouragement made the successful realization of this work possible.

For financial support I would like to acknowledge the Amelia Earhart fellowship awarded to me by Zonta in 2008.

# Curriculum Vitae

Jutta Kühn

| | |
|---|---|
| 07. September 1978 | Born in Freiburg, Germany |
| 1985 – 1989 | Primary school (Freiburg, Germany) |
| 1989 – 1998 | Secondary school (Stegen, Germany) |
| 1998 | Abitur (university-entrance diploma) |
| 09/1998 – 05/1999 | Studies in English at the University of Washington,Seattle, Washington, USA |
| 06/1999 – 09/1999 | Internship at Sick in Waldkirch, Germany |
| 10/1999 – 03/2004 | Studies in Electrical Engineering and Information Technology at the University of Karlsruhe, Germany |
| 2000 – 2002 | Vocational education + training qualifications |
| 04/2004 – 09/2004 | Research visit at National Oceanic and Atmospheric Administration in Boulder, CO, USA |
| 09/2004 – 11/2005 | Continuation of initial studies at the University of Karlsruhe with core subject "High Frequency Technology and Electronics" |
| 05/2005 | Winner of VDE Student Paper Contest in Ilmenau, Germany |
| 11/2005 | Dipl.-Ing. degree in Electrical Engineering |
| 12/2005 | Marriage with Jochen Kühn |
| Since 2006 | Scientist at the Fraunhofer Institute of Applied Solid-State Physics (IAF), Freiburg, Germany |
| 2008 | Amelia Earhart Fellowship awarded by Zonta |
| 09/2008 | Birth of daughter Eva Marie Kühn |